Bang
W
Th

James Ruppert

FORESIGHT
PUBLICATIONS

www.foresightpublications.com
www.bangernomics.com

ISBN: 978-1-9160522-5-3

More information at www.bangernomics.com

www.foresightpublications.com

A CIP catalogue record for this book is available from the British Library

Original Bangernomics Dedication

To my Father who always took me with him to buy cars and taught me so much and my beautiful Mother who tolerated a boy who filled her house with car and motorcycle parts without ever telling him off. To everyone who keeps their Banger running, whatever it is and wherever they are going.

Thanks Very Much for Saving the World

Jim Stonham got in touch to tell me that, "Bangernomics has saved me a fortune" which was generous. He shared some real world figures which made me think that perhaps I should start rewriting Bangernomics all over again. However, I am far too lazy for that. Anyway, he is at the blunt end proving that the whole concept actually works. Maybe this is a real diet after all. Also on the front line is Kenny Smith, who knows a Banger even before it fails an MoT. Meanwhile, Dee puts up with my Bangernomic nonsense random Banger purchases and provides epic support every single day. Olivia, who is better at using a spanner and looking after her car than I am.

Marion took the time to read this and point out all the mistakes.

Finally, every single reader who bothers to ask me questions, tell me off, show me what they have just bought and is generally really lovely and united by a love of the Banger.

Contents

The Bangernomics Code

Bangernomics contrasts the absurd expense of buying new, with the supreme good sense of buying used.

Bangernomics will show the mechanically bewildered how to buy a safe, durable car and most important of all save money.

Bangernomics is all about motoring at the blunt end.

Bangernomics means never having to be fussy about what you drive.

Bangernomics defines no-nonsense motoring in the new millennium.

Bangernomics is a way of life, but remember,

always beware of the dog.

Welcome to the wonderful world of buying and owning cars but with a twist, and that twist is called Bangernomics.

The original star of Bangernomics truly awful shed on wheels bought for £80 in 1990. It's a 1978 FSO 125P, which was a collection of sundry Fiat bits and pieces.

Note the sticking plasters holding on the indicator light, grime stained fluffy seat covers and an almost comprehensive tool kit. It appeared in a magazine called Buying Cars and featured the first ever use of the word Bangernomics.

Bangertroduction

We have been down the Bangernomics road a few times before, back in 1993 with the original book Bangernomics and then again in 2010 with The Bangernomics Bible and then in 2019 Bangernomics Diet. In effect what we are doing here is smashing them altogether along with countless articles and features, because what we have here is a national emergency. Actually it is rather more than that, it's an international emergency. They, that's all governments, every politician, academics, non-governmental organisations, numerous pressure groups, absolutely all of them are coming for our cars. This is our chance to save the world by running older motors. Firstly, let's recap.

Bangernomics now seems to be part of the cost saving social landscape making it into the cool kids Urban Dictionary. According to them it is "the art of buying the cheapest car/vehicle so that it does not lose value and waste money." I would argue that it is something more akin to a science, but at least the examples they provided were amusing. "Choosing an old Escort from your local garage rather than getting the latest BMW."

Even better was the example of real world usage.

"Dude, great Bangernomics there."

"Thanks, I've mastered the art."

"Drives off in a clapped out Volvo estate."

That's a Bangernomics exchange worthy of Bill and Ted's Excellent Adventure, a film that amuses and entertains, but has nothing to do with running Bangers and everything to do with time travel.

Anyway, the point of banging on yet again is that Bangernomics still needs some explaining and possibly a serious expansion of what it actually means. Not that I am the beginning and end of the definition, it can mean different things to different drivers and car owners and always will. The trouble is that some just think that Bangernomics is buying a Banger for a few quid and then running it into the ground, or just leaving it on the hard shoulder for the emergency services to deal with. I'd like to think that Bangernomics is actually much more subtle and clever than that. Certainly the stories I've heard from Bangernomicists have been rather inspirational and that is reflected in this attempt to save the world.

Also, the word 'Banger' needs to be reclaimed. There are some sweary and quite silly words being used to describe older cars and I think that Banger has a certain purity and purpose about it.

Back in the 1990s the follow-up to Bangernomics was going to be called Carobics and that dealt with keeping a Banger on the road for as long as possible. Not quite sure why that book didn't happen, it was the automotive equivalent of aerobics, you know, keep yourself fit and you'll live forever. Well, for me Bangernomics is all about that too and making a Banger last longer. Indeed, there were quite a few books available in America on that theme, including 'Drive it Forever' by Robert Sikorsky and "Why Trade It In?' by George and Suzanne Fremon. I think that there is certainly a book to be written by a proper mechanic on how to do that. Certainly I will touch on simple ways to make your older car last longer. There simply is no point running a disposable Banger. Obviously some models do have an expiry date on them because repairs may become uneconomic, or there is a catastrophic failure. You may also simply be bored with it. When it comes to price, there isn't a strict limit, or level because everyone's budget is relative. A Bangernomic car has to be one that you can afford, without any borrowing, but there is no point putting a figure on it. Figures, or rather, values, change. Inflation happens and suddenly a book looks very dated. I think we are all grown up enough to know what a cheap car actually is, but it always has to be affordable.

The great news is that *Bangernomics* will be giving you, dear reader, just the excuse you need to run a sports, luxury or classic car within the Bangernomics ethos. There is more to Bangernomics than you could have imagined, but I will be keeping it all dead simple.

The simple truth is that any idiot can buy a used car and every year many do. Also any idiot can run a used car into the ground, but that wouldn't be very clever at all. Often the most idiotic thing that anyone can do is buy a brand-new car and then sell it a few years later at a crippling loss. But even more stupid would be not to learn from that and go back into the showroom and start the whole depreciation process all over again. Yet millions do and for many it doesn't have to be like that.

Once upon a time you bought a car to do a job and then ran it until the vehicle outlived its usefulness, or failed mechanically, rather than making it part of some sort of lifestyle package. What I mean is, there has been a growing trend for car buyers to convince themselves that it is a matter of urgency to replace their brand new car every three years or sometimes less with another new car.

The great news is that Bangernomics allows you to have your car shaped cake and actually drive it. Because if you love cars then you can buy a dull but reliable domestic appliance on wheels for the weekday commute, whilst in the meantime save enough to buy something much more interesting for the weekend. Or of course you could spend what you save on something flippant like food, or the mortgage. Bangernomics will undoubtedly save money for the motorist who just wants to run a car without a breaking the bank and does not give a stuff about acceleration and top speed stats, colour, or anything else that is utterly irrelevant to them.

I wrote Bangernomics back in 1992 when airbags were just bags with air in them, people carriers were still vans with windows and air conditioning meant opening the windows. That has all changed as have the values of used cars and what counts as a Banger. Back in '92 even a reasonable amount of money would only get you a scruffy, possibly unreliable, borderline roadworthy vehicle. Now even a modest budget gives you the pick of a wide variety of vehicles that are often in decent condition with full service histories. The new car culture has changed so much over recent years that a Banger can now be a car that is just out of manufacturer's warranty. That's because easy and cheap new car finance, the steady reduction in new car retail prices and changing regulations regarding the sales of cars have all helped to bring about a fall in second hand values around the world. Now even a five to seven year old car can be unwanted and almost worthless, which is why Bangernomics, as a method of buying and running cars, is still so relevant.

The principles of buying a used car are simple enough, which is where Bangernomics comes in. I believe we can all make a common sense assessment of an older vehicle without any mechanical knowledge at all. That's why I have decided to strip out most of the technicalities, as there is no point droning on about complicated mechanical failure, when simply stating the obvious would be better. So if something is noisy it is probably broken or about to become even more broken.

Changes from previous Bangernomics editions means that I have deleted any facts and information, which will inevitably date. You can look up a lot of detailed information about specific cars, road taxes and all that on the world wide Interweb, which will still exist in some form or other and may well be hard wired into our brains by 2075, or earlier. So when your grandchildren buy a potato powered hover plane sometime in the future, the same basic principles will apply. I believe that you will only ever need one book on how to buy a car and Bangernomics will Save the World, is it.

Indeed Bangernomics is now truly international as the principles of buying a car and running it for not very much are practised absolutely everywhere, especially in rural communities. The French are not running around in no hubcap, matt paintwork, hatchbacks for fun, they do it because it is the cost effective thing to do. Go to India and watch them repair cars rather brilliantly in the street. Take Cuba, the residents of that country had no alternative but to practise an extreme form of keeping old American Bangers running in perpetuity, or at least in Havana. Their resourcefulness has been a fantastic example to us all and they certainly won't care that we have given what they do a name, because there has been no alternative, due to their demonically awful political system.

If you are in Cuba or outside the UK, I will try and avoid any obscure references to the UK car tax system, although similar measures are being applied worldwide. So there might be references to the annual roadworthiness check called the MOT. However, I will keep the information general and straightforward in the interests of international understanding and co-operation, but also so that I don't have to rewrite Bangernomics yet again.

*And then there is the question of support. I've been doing that for decades with a quill pen, semaphore, typewriters and sometimes just by shouting. That may not be good enough anymore, so as an act of Bangernomic good faith I promise to provide you with lifetime support. Initially it will be my lifetime, however long that turns out to be, although I will try and sort something out in the far-flung future with relatives and creditors. So while emails and the Internet exists you can contact me (for as long I exist anyway) for biased advice through Bangernomics.com.

Also what we want to encourage with the Bangernomics is to get people actually looking at cars in the metal again. You might see some pictures on a screen, maybe even a film, but there is never any substitute for actually seeing and driving the car. There is also nothing better than actually meeting the person selling the car. If you don't trust or like them, how can you ever trust the car?

Don't forget that at all times you are in safe, if slightly shaky hands. Because when it comes to buying cars I have made every possible mistake. I hope you can learn from them.

Let's explain the cover a bit, The picture was taken almost 25 years ago in 1997 at the German launch of the Volkswagen Golf Mark 4. In theory that particular Golf whether it was a 1.9 diesel or 1.6 Petrol should still be running around today as the ultimate Banger. Those Mark 4s were very well built, pretty rust resistant, nice to drive and practical. The V5 was fun. The GTI was a slight return to form. And Helga? Well, I was in Wolfsburg, the home of VW, partly because I was writing a book on Golfs and also on behalf of Golf Xtreme magazine. That was an enthusiast orientated publication which had a slight eye for the ladies and it seemed like a good idea at the time to politely ask the permission of one of the hostesses. Helga really was more than happy to pose, she obviously spoke English and found it really funny that I wanted to take her picture.

Back to the book, don't buy a Dog, buy the Bangernomics Will Save The World. Oh you already have, thanks.

James Ruppert 2021

Free Lifetime Support*: contact the experts (still only me) at Bangernomics.com for free advice forever on what to buy and where to buy it. *Subject to Terms and Conditions which are lightly touched on in the introduction. Just go to Bangernomics.com to check the latest information.

Publisher's Note: Whilst the Publisher and Author (same person) have made every effort to ensure the accuracy of the information contained within this book. No liability can be accepted for any car purchase made. In the case of any doubt regarding the condition or safety of the proposed purchase, readers should seek professional advice and not proceed with any purchase where doubt remains. Get a professional involved because we can't be there to hold your hand.

1. The World does not need Saving *and neither does the Universe...*

...Not from us older car owners, fanciers or Bangernomicists anyway. What we need to do here is get all of our 'used cars are better for the environment' argument ducks in a row. Although there are debates to be had and there does seem to be a big government level agenda to get all conventional cars off the road. So let's keep it to old cars to prove that what we do in the name of Bangernomics is good for everyone.

The European Federation for Transport and Environment in a report published in 2018, stated that in order to meet the 2050 Paris climate commitments cars and vans must be entirely decarbonised. This required ending sales of cars with an internal combustion engine by 2035. As we know this deadline has been brought forward by the UK Government to 2030. The truly terrifying sentence in that report was, "Such a transformation requires wholesale changes, not only to vehicles but also how they are owned, taxed and driven." That and much more is certainly occurring right now. Bangernomics is one way of fighting back.

According to some environmentalists, even Banger ownership is irresponsible because of the pollution they cause. However, a Bangernomic approach to motoring amounts to recycling. As the natural resources and energy used in building a new car is phenomenal, prolonging the life and disposing responsibly of a Banger is very green. Also see a reproduced interview at the end of this book which I did with Charles Ware a few years ago about his real word superhuman effort to get us practising what he called in the '70s, 'sustainable motoring'.

Right then, let's get on with laying out the reasons why the world, never mind the universe, will be saved (certainly saving an awful lot of money) if only everyone who owned a car decided to apply some Bangernomics to their motoring lives.

Bangernomics parked next a brand new Car

It is hard to escape the simple fact that the natural resources and energy used in building a new car is huge. Around 5.6 metric tonnes of carbon would be used during manufacture (more details further on). That is a rough figure covering energy for assembling and painting on the production line, the logistics (fuel used by lorries), including the tools, plus raw materials consumed.

As a result, prolonging the life of and later disposing responsibly of a Banger is very green indeed. Permission to feel smug if you are doing the Bangernomics thing. Indeed the mainstream intelligensia are only now catching up with us. In the updated 'How Bad are Bananas? The Carbon Footprint of Everything, Mike Berners-Lee argues that in most circumstances keeping an old Banger running is very good indeed. Meanwhile, Electricity Cars are off the so-called Carbon Footprint chart. I am paraphrasing, but essentially we Banfernomicists are on the right side of history.

In recent years, the gradual, real world fall of new car prices, cheap finance and a culture of increasingly rapid automotive obsolescence has meant that cars past their fifth birthday can now cost marginal amounts of money and qualify as a high quality Banger.

The point is an old car has already been built many years before and does not have to be made again. The massive initial energy has been expended, maybe a decade before. Older car owners are in the clear when it comes to environmental harm compared to a new car buyer. The trouble is that you really do have to believe that $CO2$ is bad instead of a naturally occurring gas to think it is harmful. However, that is the politically driven system of measure which is now in place. Unfortunately we are obliged to use it when calculating day to day emissions..

So If you want to work it all out for yourself one litre of diesel corresponds to 2.68kg of $CO2$ and 2.31kg for petrol. Simply multiply the number of litres used over a set period, (a day a week, month or year) by 2.68 or 2.31 to get the total kilograms of CO_2. I would keep it to the running costs because the initial energy use for production is almost ancient history and for older cars is very difficult to calculate anyway.

Getting back to new cars, interestingly, The Guardian of all newspapers sought to explain what the carbon footprint of a new car actually is, and did a half decent job of explaining how difficult it is getting to the bottom of it all.

So for a basic City friendly Citroen C1 Urban Ride it would be 6 tonnes CO_2. Moving on up to a family and business sized Ford Mondeo in middle order Titanium specification produces 17 tonnes CO_2. Then the worst case CO_2 scenario is going to be one of those pure evil incarnate sports utility vehicles. A Land Rover Discovery SD4 SE is going to produce a whopping 35 tonnes of CO_2. The Guardian touched on all the complexities of manufacture when it comes to calculating the carbon footprint. It is an almost impossible undertaking.

However, taking the input-output approach is to deconstruct the known total emissions of the world or a country into different industries and sectors, in the process taking account of how each industry consumes the goods and services of all the others.

All the gas and electricity used by manufacturers, including the component manufacturers, accounts for less than 12% of the total. The rest is spread across everything from mining (33%), rubber manufacture (3%) and the manufacture of tools and machines (5%) through to business travel and stationary for car company employees. Then divide all that by the total amount of money spent on new cars, the Guardian managed to reach a footprint of 720kg CO_2 per £1000 spent. As a result, spending £24K on a new car, which isn't difficult at the time of writing, amounts to 17 tonnes of CO_2, which the Guardian reckoned was equivalent to three years' worth of gas and electricity used in the average home.

That meant the embodied emissions of a car

typically rival the exhaust pipe emissions over its entire lifetime. Indeed, for each mile driven, the emissions from the manufacture of a Land Rover Discovery that ends up breaking down and being taken apart for spares after 100,000 miles may be as much as four times higher than the tailpipe emissions of a Citroen C1.

Here's the thing, The Guardian said "unless you do very high mileage or have a real gas-guzzler, it generally makes sense to keep your old car for as long as it is reliable – and to look after it carefully to extend its life as long as possible. If you make a car last to 200,000 miles rather than 100,000, then the emissions for each mile the car does in its lifetime may drop by as much as 50%, as a result of getting more distance out of the initial manufacturing emissions."

Could not have put it better myself, well I have been for over thirty years. In the interests of balance (yes against the Guardian), Green Car Reports disagrees. According to them, if the older car does just 30 mpg, and you drive 15,000 miles per year, but the newer car does 40 mpg. Then during those 15,000 miles, your old car will use 500 gallons of gas (petrol) and the newer car: 375 gallons. That 30 mpg is also about 290 grams per mile of CO2; 40 mpg works out as 217g for every mile. Over the course of the year, the old car will emit 4.35 tons of CO2, the newer car 3.25 tons. But if you've been running the older car for 10 years, it has emitted over 43 tons of CO2 before the new car has even been built. There is of course lots that you can pick apart in that. Not least some woeful American mpg figures and going on about previous CO2 emissions from back in the old days, whu=ich is utterly irrelevant.

Bangernomics versus the Green Agenda

Fundamentally it is the demonisation of CO2 which has led to the current predicament where any mechanical activity is perceived as damaging to the planet. That's the agenda adopted by so many governments especially in Britain. In June 2019, it became the first major country to legislate for a net-zero target for carbon emissions by 2050. Here is a brief word salad extract from the announcement: "Government needs to shape policies and regulations to create a market environment that increases consumer and business demand for low-carbon solutions and encourages sustainable private sector investment decisions. Policies across all areas of government interact to influence the transition of the whole economy towards net zero."

The government want to stop us driving and even heating our homes in a conventional manner, be it gas or oil fired heating and even wood burners. Most of all Net Zero does not make any sense if just one country, Britain, actually does it. Any supposed benefit is wiped out buy just one dictatorship such as China continuing on as before. Indeed, China approved almost 10 gigawatts of new coal-fired power generation projects in the first quarter of 2020, almost equivalent to all the power stations they built in 2019. Britain has already exported any CO2 emissions to China and elsewhere in the world by becoming a consuming rather than manufacturing nation. Reducing the UK's less than 1% contribution to global man-made carbon dioxide emissions to net zero by 2050 would reduce global average temperature by a pretty much unmeasurable 0.014C by 2100. Similarly, even if all the countries signed up to the Paris Agreement keep their emissions reductions promises, the reduction in carbon dioxide emissions for 2016 - 2100 achieves only 1% of the reductions required for the stated 1.5C target.

This whole agenda is all about controlling what we do, rather than the climate. You can't control the climate. Mother nature is more more powerful than you will ever be and if anything controls what happens on earth it is ultimately solar activity. The so-called scientific community has flip flopped from predicting an ice age to heading down the global warming route before calling anything and everything climate change. Right now it does seem as though the weather is cooling which interferes with the prevailing hot house narrative. In 2013, the Intergovernmental Panel on Climate Change conducted a thorough survey of the projected impacts of global warming in the context of the other trends in economic growth and development. They concluded that the effects will be real, but relatively small and manageable.

Yes the climate changes, always has but there is very little you can do about it. Pollution? Absolutely, clean up your act and in motoring terms recycle old cars through Bangernomics, then you really are doing your bit.

Bangernomics and Battery Cars This might be a good time to ponder just how green an electricity vehicle is. What we have been sold is a car which can be filled up with power for just a couple of pence, causes zero pollution, is whisper quiet and needs minimal maintenance. A purely electric car would appear to be environmentally and practicality the answer to all our motoring needs. This is certainly how electricity cars have been promoted so far.

There are different kinds of electricity vehicles. From the purely electric to the hybrid which gives a conventional engine some battery assistance, plug in hybrid and even a range extender. To keep it simple we will stick to the purely electricity vehicle, also referred to as an EV to make it sound all very futuristic. Emissions may appear to be zero, but recharging the batteries takes electricity. This electricity can come from fossil fuel burning, pollution heavy, coal/oil/gas fired power stations. However, invest in some solar panels, or erect a windmill in your garden and you could recharge with a reasonably clear conscience. Except that everything has a cost as we have seen when it comes to simply building and running a conventionally engined car.

According to the laws of chemistry an electric car uses the same fuel as a petrol car, but burning much less of it. Although the greenhouse gases would be harmful wherever they occur, it could be argued that if the emissions happen at a power station, rather than at the roadside, it might be better for the nation's health. Mind you building wind turbines is not emissions free, plus those solar panels didn't get from China on their own.

Mostly though, batteries are expensive and don't

last forever. Makers of electric cars sell their models with a warranty for the battery, which will lose capacity over time. For instance, in the case of the Nissan Leaf, it is covered for five years or 60,000 miles, rising to eight years or 100,000 miles for the version with a bigger battery. Indeed, most manufacturers will sell you a car but lease the batteries. Replacing those batteries when they are below 80% efficient after eight years or so, will cost thousands and that does not seem particularly sustainable.

Then there is the very

real fact that making the batteries is not at all eco friendly because of the huge quantities of rare earth materials that are required to make them such as Neodymium, Lithium ion as well as Cobalt and Copper. The quantities required leads to environmentally disruptive mining and is sometimes by children in the Central African Republic. So surely that's a fossil fuel nightmare and currently phenomenally difficult to recycle. Please see my book Demotorized for the full horrific details. You don't have listen to little old me, it has been estimated by engineering company, Ricardo, that the production of a petrol car creates emissions amounting to 5.6 tonnes of CO2. For an electric car that figure is around 8.8 tonnes and half of that is making the batteries. However, over the lifetime of the vehicle an electric car would be responsible for 80% of the emissions of the petrol car. Even so, running a conventional petrol or diesel car is actually the best approach for responsible car ownership, as it will last longer than an EV and be cheaper to buy. There is also the rise of Hydrogen and also synthetic fuels. We don't have the time, the space or inclination to investigate where vehicle propulsion is going. The simple truth is that right now you are simply virtue signaling if you buy a car with a great big battery just to make a point. Oh, and just in case you wondered, Battery, or Electricity cars can't ever be Bangers.

Bangernomics versus the New World Motoring Order and the new car sales ban in 2030.

Quite clearly there is an agenda to get us all out of our cars, make us feel guilty about the act of vehicle ownership and remove any pleasure from simply driving. That's what the United Nations want in their Agenda 2030 and it seems to be that the British government are fully on board with this attack on personal transport and removing freedom of choice. If this trend continues then car ownership will become the privilege of the very rich and the entitled (for instance members of parliament). The rest of us will have to wait for a bus, use our bikes or walk. Very simply, they (governments, non governmental organisations, the United Nations, the green lobby) want to restrict our freedom of movement.

On the face of it why should any driver committed to Bangernomics care? After all, driving an older car will mean a Bangernomicist is in the clear. Anyway the government decision in November 2020 to bring forward the deadline to ban the sale of new petrol and diesel vehicles by ten years was a bit of a shocker. They laid out plans to phase out the sale of new cars and vans run on traditional fuels by 2030. Step Two will see all new cars and vans be zero emission at the exhaust pipe from 2035 when they ban the sale of Hybrid vehicles. Great. I did see that one coming in my book Demotorized, but the surprising thing is how little push back there has been against this. Car magazines, motoring organisations, journalists, essentially the whole car based establishment said nothing much about it at all. They never will because there are advertising contracts, loan cars and new technology to get excited about. Never mind that efficient, economical and practical cars are being replaced by what are vehicles that represent a very real compromise. So it is up to us to do something and fight back.

Of course there is always the chance that this ban will not happen, because governments often cock things up. So it may be delayed or modified not least because there is a chance that the National Grid might struggle when millions of drivers switch to electric. Or possibly when a few of us plug the kettle for a brew there's an almighty power cut. Who gets priority? Will we be relying on next door's Tesla to store some 'lecky so we can have the lights on? The Government had made available £1.3 billion of taxpayers money to accelerate the roll-out of charge points for electric vehicles in homes, streets across the UK and on motorways across England. This is unlikely to be nearly enough money to make this happen, so they will probably take much more.

We can't rely on any of the motoring organisations standing up for us. They don't care about you and just want to sell you insurance and possibly lecture you about something vaguely motoring related. The motoring media continue to be a massive disappointment all round. There are no national campaigns to halt this huge change in the way we own, buy and drive cars. Maybe if there is an election you could vote some other party that might change this policy? Well actually no, they are all on board with the same net-zero, punish the poor motorist forever, agenda. Unless you plan on starting a pro driver political organisation that we can all get behind, the simple truth is that nothing is going to stop this nightmare happening. Telling your local MP politely, but firmly that you won't vote for them ever again might concentrate their tiny minds, but only if tens of thousands do it. Otherwise they will continue to take our votes for granted.

Which brings us back to the original proposition that those who abide by Bangernomics won't be buying a brand new car in 2030. Maybe not, the point is that the choice will be taken away. Many Banger fans have run brand new cars for three years because of work commitments and then had maybe a spare Bangernomic car, or something cheap and cheerful, or just fun for the weekend. Never mind though because cars with old petrol and diesel engines can last decades, so we will be fine. Except that now the authorities and the politicians have the taste for all this banning malarkey it would be easy enough to put a best before date on petrol and diesel. Not just ban the vehicles, but the fuel that goes in the tank. That has now happened with the introduction of E10 fuel which cannot be used in the majority of cars manufactured before 2002. All of this is being done on the basis that it is for our own good. This Net Zero, Green Agenda, Extinction Rebellion inspired activity and agenda is just pure misguided evil.

Unfortunately we need to proceed on the basis that the authorities are coming for our cars in 2030 and thereafter. Clearly no one in authority is ever going to help us. For that reason we will have to help ourselves. Here is a half baked nine point plan that may get you into some trouble, or it could turn you into a hero. The thing is, it doesn't just need you, but your family, friends and work colleagues. There are millions of us drivers, never mind a couple of Bangernomic nutters, like us. Here are some suggestions.

1. Bangernomics Network It really is up to us to get together with others who also want to keep their cars going for longer. We can do this in a number of ways. Joining an owners club, which for an older car is easy and probably essential. You are instantly plugged into companies that can supply hard to get parts and members who have lots of expertise and opinions of course. If you do have a rare or very new model, then maybe you have to take the initiative and sort something out yourself. I will certainly do as much as I can through the Bangernomics website. I would love to issue a membership cards again and maybe I will. At the very least we can all put each other in contact with garages and other organisations who will help. They could be a good independent garage, or maybe a company offering a whole life warranty, or something that will just make our motoring lives easier and make our Bangers last longer.

2. Join this lot. I know I said the AA and RAC are useless at sticking up for us, but the Alliance of British Drivers and the Fair Fuel Campaign are at least a million times better. They might even get the job done. Given time. There isn't anyone else trying to put motorists first. Both those organisations may not care a jot about Bangernomics, but they are committed to the continuation of private affordable mobility. You know, motoring.

3. The Classic Option There is just a chance that the people in charge could let one group of motors slip below their 'Ban all Cars' radar. The demographics for owners of older cars is rather like the motors themselves, old and quite often decrepit. It is a declining market, but maybe this is a way to get new recruits. Bangernomics should prove to youngsters that an older, simpler, historically significant car is the way forward. These may be allowed to exist for the moment, but there is always the chance that classics could be regarded as recreational vehicles and owners only allowed out once a week or when there is an R in the month. You just never know, but at the moment vehicles over 40 years old are exempt from tax and also allowed into Ultra Low Emission Zones. These old Bangers that are now classics are possibly too rare to bother the authorities, for the time being.

3. Importation Situation Registering brand new internal combustion engined cars may not be allowed from 2030, but if you bought a car outside of the UK? Technically the year it was built could scupper that proposal, but if it was a private transaction and not a business it may be allowed. Maybe you might have to tolerate a European left hand drive specification. Then again there are right hand drive Japanese cars. Vehicles from overseas had to be driven for a period in their country of origin, usually six months. It is a thought and this option will be worth investigating nearer the ban.

4. Fuel for Thought Any government and indeed business that wants to control how you behave and what you drive, can outlaw or just stop the supply of petrol and diesel. It is an idea to consider alternative. Liquid Petroleum Gas has been an option for years and half the price of fuel it is a bi-product of. It is technically possible to run an older combustion engine on alcohol. Making alcohol without a licence is not legal of course, but it can be done. There are books available on how to do this. Once upon a time there were steam vehicles, not terribly practical, but we should start thinking about alternatives to the alternatives. Worth exploring as it may be the only way to keep your adapted Banger going.

5. Always Campaign I don't think many car owners have any proper understanding of what the 2030 ban actually means to their own mobility and freedom of choice. The ban seems a very long way off to most people. It is our duty to bring up what this really means for their future and bring these talking points up with friends, family and just anyone who engages with us in a motoring environment or context. You know, a queue at a petrol station. (I wrote this before queues and fuel poverty suddenly reappeared from the 1970s).

6. Build your own car. Far fetched, but don't forget that there has been a whole car culture and micro industry based on blokes in sheds building cars from kits or just a random collection of parts. Indeed, this is how the motor industry started. Think of something different, you might just make a large fortune. When it comes to long term vehicle ownership this may be the best way forward with like minded, possibly more mechanically inclined friends and acquaintances. Maybe the future might be making a trike or something which leads us onto...

7. Bikernomics. I don't know very much about motorcycles, mopeds or scooters, but they do offer an alternative, a back up and a cheap to run option if ever the main Bangernomics motor goes down. Actually the traditionally engined bike has been as much under attack by environmentalists and governments as the car. That makes bikers are our allies in all this. Anything is better than those terrible electric scooters that the authorities want us all to drive around town and for that all important 'last mile'. There is a chance that the motorcycle might be left alone, but they are as much under threat as cars.

8. Technophobia Essentially by going down the classic route you are dodging the worst of the 'connected cars' that the authorities want us all to drive. It will make their lives so much easier and simpler if they know where we are all the time and then even turn off what we are driving. That is so simple with moderns and electrics. At the very least give up your Smart Phone. I know it is indispensable and makes your life complete, but really try a bit of cold phone turkey. Get a dumb one, a satellite one. Use a pay phone. Cut yourself off from the grid. This might be impossible. At the very least it would be interesting to try and discover what technology you can do without.

9. Become Disobedient Drivers The is the last resort of course. I could not possibly condone any kind of bad behaviour at all. However, if driving is outlawed or certainly proscribed, what could a driver do? Guerilla tactics would have to be the way forward. Mass disobedience, for example by all of us liked minded Bangernomicists going for a drive to the coast at the same time. Need a green numberplate to comply? Then make sure the car has one, by possibly applying some coloured sticky tape. This movement does not need one leader, or figurehead to lead the fightback. That's because every driver is now a soldier in the fight. Arrest one and there is another waiting to jump into the driver's seat. That's the theory anyway. The reality has been so far that the majority of us struggle to do anything. Hardly anyone signed a petition asking the British Parliament to think again about the 2030 new car ban. We can dream about fighting back, but we really must act.

2.Big Bangernomics Theory

Let's go through some basic questions and answers...

Want to cut motoring costs? Do you want to buy a good used car? Answer yes to either of those questions and it is a certainty that you need Bangernomics in your life. Just in case you haven't got the message yet, Bangernomics contrasts the absurd expense of buying new, with the supreme good sense of buying used. Bangernomics will de-mystify the art of buying a used car, help you save money and promote safe motoring; it's as simple as that. The worldwide used car market is massive, far more used cars are sold than new, so it is in every motorist's interest to be completely clued up about the Bangernomic lifestyle. However, if you still have lingering doubts, lets deal with some of the most common Bangernomic FAQs as theoretically posed by you.

Does Bangernomics Save Money?
In both the short and long term a Bangernomic approach to car ownership makes sound financial sense. The truth is that purchasing a new car involves immediate financial losses. Before it has turned a wheel the government imposes all sorts of taxes and that accounts for a large proportion of the purchase price, which has to be written off. Add to that the dealer's charge to put the car on the road and you can start to feel rather unwell, but the bad news does not end there. Depreciation takes an unhealthy bite out of the car's value too. Depreciation is at its worst in the first year, but it also keeps on eating away during the warranty period and doesn't stop there. With a Banger, or at least a slightly older car all the previous owners have taken the brunt of depreciation and high initial running costs. That leaves the clever Banger buyer with a vehicle that is unlikely to drop in value by very much, having reached rock bottom, or at least close to it. A few years down the line a Banger owner may

even get a significant amount of their money back when it turns into a desirable classic.

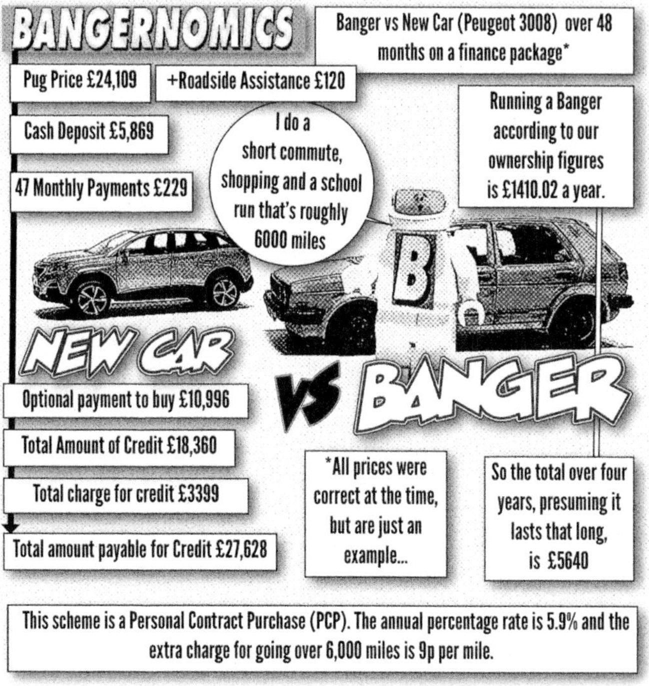

But aren't Bangernomic costs the same as a new car?
No. Let's take servicing. Some older cars can be simpler to work on and therefore cheaper to maintain, especially if you carry out some basic checks and repairs yourself. Even if you do buy a fairly complicated and more modern car there are specialists who can look after it but at much more friendly labour rates. There are also some excellent independent local garages which can often solve problems, rather than simply fit another costly part as main dealers sometimes do. Incidentally, so much information is available

online, especially videos showing us how to do some of the simplest and most complicated jobs on your Banger. There are also alternative sources for parts rather than paying high main dealer prices. Bangernomics will even show the mechanically cack-handed how to do the minimum maintenance for maximum mileage. In some cases a Bangernomic vehicle only merits the lowest insurance cover. Depending on the age and type of Banger a dent might even improve its looks so an expensive visit to the body shop is unnecessary. Owning an older car means that there is also little reason to worry about the bump and grind of close quarter city traffic and no lost sleep that a vandal might add another scratch to the already weathered paintwork. Of course you should be conscientious about servicing, maintenance and general cosmetics, but you should never be obsessed or worried about getting very minor damage fixed. So Bangernomics helps you take a little bit of stress out of owning and driving a car.

What is a Banger?

Difficult one that. To some people a car becomes a Banger the first time a windscreen gets a fly spattered across it, others may think that it has to be at least a decade old, or when it gets its first stone chip. However, to the Bangernomicist, it is a car that is attractively cheap, which still has some mileage left in it. Ideally it will be a car that you can afford, rather than one which requires a big loan to buy. Bangernomic cars are mostly out of manufacturers warranty and might even have the sort of mileage (100,000 miles plus) that puts some buyers off. As I said in the introduction, things have changed and just about any car that is outside of the manufacturers' warranty can now potentially be classed as a Banger. There is a bit more detail on what is and isn't a Banger coming up. You should be pleased that the definition is now so wide and embraces all sorts of exciting models.

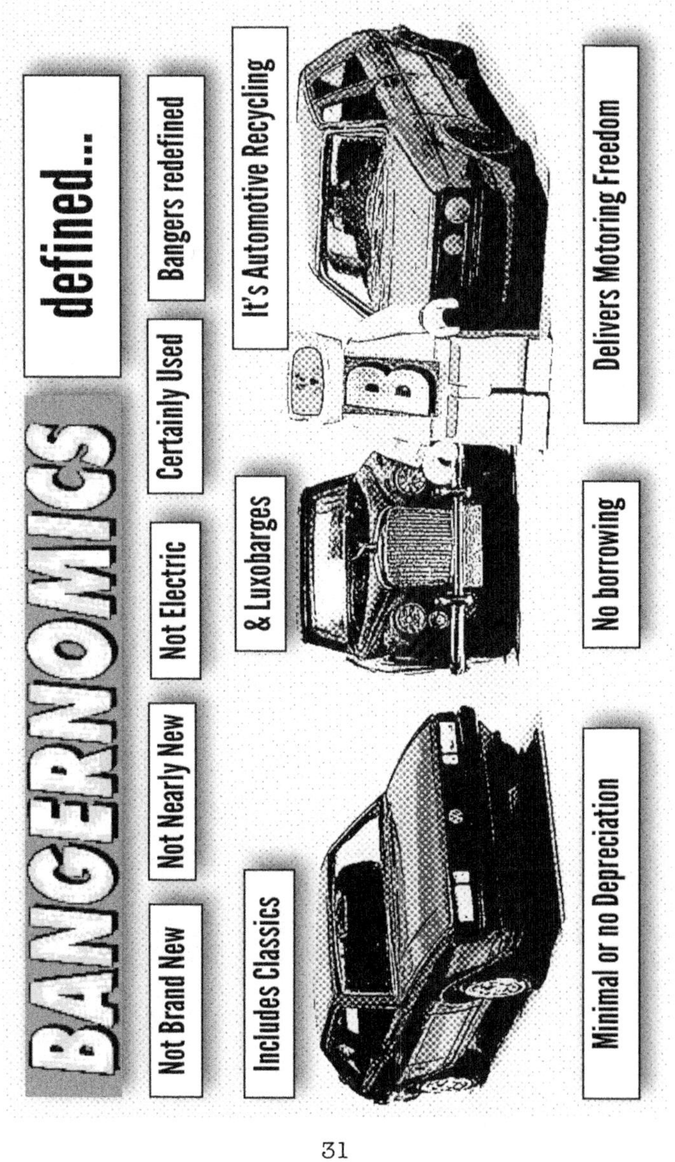
BANGERNOMICS defined...
Not Brand New
Not Nearly New
Not Electric
Certainly Used
Bangers redefined
Includes Classics
& Luxobarges
It's Automotive Recycling
Minimal or no Depreciation
No borrowing
Delivers Motoring Freedom

Who won't be suited to Bangernomics?
Snobs who feel obliged keep up with the Motoring Joneses, and must have the latest, most expensive model and then constantly remind everyone how much they paid for it. Basically anyone with more money than sense won't be Bangernomically inclined. Quite a small minority then. Bangernomics is for everyone who drives.

Who benefits from Bangernomics?
Everyone who currently owns a car, or needs to buy one can immediately feel the positive effects of Bangernomics where it matters most, in the pocket. Of course, lower motoring costs are the goal of all private owners who do not have the safety net of a company car. However, even those who have a relatively new model as first-string transport would do well to think about having a usable second car for shopping, or commuting. Why buy a brand new car simply to drive a few miles to the train station?

Is Bangernomics Green?
We have covered that, but here is a little recap, essentially this is what I wrote almost 30 years ago... So, for you environmentally concerned little greenies out there, it is and it isn't. When you consider the amount of pollution that engines cause, car ownership of any kind could be viewed as irresponsible by extreme car haters. However, a Bangernomic approach means that a car is effectively recycled, rather than abandoned. The natural resources and energy used to make a new car is phenomenal, which makes prolonging the useful life of a car and then disposing of it responsibly decidedly 'green'. So put your bike in the shed and get motoring.

Is Bangernomics Risky?
Of course it is. Buying a used car brings with it some extra concerns and there is always a degree of luck involved. However, as Bangernomics shows it is possible, by just using common sense and no specialist technical knowledge, to avoid the worst pitfalls. Indeed, purchasing a new car does

not necessarily mean hassle free motoring. It may break down or prove to be unreliable leaving the buyer with precious few legal remedies. Bangernomics is all about reducing the risk and putting a satisfied smile on your face.

The Bangernomics Bottom Line

If you can afford a brand new car and that's what you want then fine. Just don't start crying when you sell it after a few years and find that it is worth buttons. However, keep it for a decade or two and get full value from it and no one, least of all a Bangernomicist, can argue with you. In previous editions of this book I have compared and contrasted different old, new and slightly old vehicles, but that's something I'd prefer you to do. You will learn a huge amount, especially once you have decided on the type of car and particular model that you think you want. So look at new prices, used values, servicing costs, depreciation and do the arithmetic.

What can we do?

Frighten Ourselves: Contact some car dealers and ask them how much their new cars cost. However, be on guard and don't get tempted to buy new and slap down a deposit. That would be a disaster. If what the salesman says does not scare you, do your own bottom line calculations for a basket of similar cars and write down what they cost new, the servicing costs and their value after three years.

Dream on: What would really like to do or buy? A holiday, a self-improving course in flower arranging or Karate, maybe a set of golf clubs, or a kitchen extension? Get quotes for all those things that could genuinely enhance your life. Many of those things are probably a lot less than the three-year depreciation figure for a brand new car.

3. Bangers & Cash:
Budgeting for a Banger

This is serious. Get it wrong and you will be in financial trouble and that's means missing the whole point of Bangernomics as a guide for your motoring life.

The true Bangernomicist tips up their piggy bank and shakes out the remaining silver and copper coins to find out what their used car budget really is. If you don't know how much you can afford it will certainly all end in tears. In the long term borrowing money ends up costing you money because of interest charges. Yes there are people who don't realise that, or just don't care. That means you must carry out some basic budgeting, not only to arrive at a sum which goes towards the purchase, but also a weekly allowance that means you can afford to run the car. This is basic housekeeping and if you have never sat down and worked out what you can afford, now is the time to start. And so you can't wriggle out of your budgeting commitments it might be a good idea to write the figures in this book in ink. Don't even think about changing the budget figures unless you suddenly come into a lot of money courtesy of a great new job, lottery win or an inheritance.

Anyway here's a Bit about Borrowing which Bangernomicists don't do of course. Drivers who don't have the money to buy a new or replacement vehicle should probably save up, that's the Bangernomics approach. However, on that the basis that real life isn't always so simple and the demands of family, or work means that a reliable Banger is a priority, let's take a quick peek into the world of borrowing money to see how scary it actually is. For the buyer who spots a Banger in the classified ads could equip themselves with a personal loan. All they need to do is pop along to their bank, building society or a finance company. Of course, credit ratings must be good in order to secure the money. Ideally, shop around for the best possible deal as the annual percentage rate will vary. It is easy enough to avoid loan sharks and sticking to the lenders you know best.

It is car dealers who have the most creative borrowing solutions when it comes to a replacement vehicle. Hire purchase where the loan is secured against the car. Ownership does not pass until the last payment has been made. The dealer hopes that a low deposit and flexible repayment terms from twelve to sixty months (that's five years) is all the convincing you need to go ahead and sign up with them. Car dealers and finance companies new favourite car use scheme has been the Personal Contract Plans (PCP). The monthly payments are lower and the ownership of the vehicle stays with the finance company. The good news is that there are options once the scheme is over. Simply pay the resale price and keep the car, trade it in and take out another PCP, or just hand the car back then walk away. Finally, for complete motoring peace of mind there is personal leasing. Here, the buyer pays the dealer a fixed monthly amount for the use of a car, with servicing and maintenance included. However, the mileage must not exceed a specified limit and at the end of the lease agreement the vehicle goes back to the dealer. Overall, this is a more expensive way of running a car, from a high initial deposit to increased monthly payments for servicing and maintenance.

Motorists need to make the choice between being frugal, responsible and living within their means with Bangernomics, or having the latest, connected, hybrid Sports Utility Vehicle thing. If it is the latter then they must compare schemes, lenders and different annual percentage rates, which will reveal the true cost of the loan over the whole period.

Borrowers must be aware of additional costs that penalise early payment, excessive mileage and damage. They should also consider taking out GAP cover which will pay out in full if the car is a complete write off and most importantly pay the outstanding finance on the car, even if that amounts to more than the model's value.

The choice is yours, do you believe in Bangernomics, or are you just a borrower?

GET BUDGETING: The logic is simple, how much do you want to spend? £_______ (write it down) How much do you have saved? £_______ (write it down). If there is a shortfall, then think again. If there is a surplus, is that enough to constitute 'savings' given your lifestyle and requirements? If you can't afford it, then maybe you should forget about buying for a few months. The Bangernomics approach is to do a bit of overtime, save up and use your own hard-earned cash to qualify for the Bangernomics course. If nothing else, it concentrates your mind wonderfully on making a purchase that you won't regret. Even if you make a mistake and buy a dud then at least you won't be paying for it over the next 12, 24 or 36 months or more as you would with a loan. Once you have decided to spend a certain amount on the car write it down in this book in ink. As you will know or soon discover, the big problem when buying a car is that you can get carried away and end up spending more than you have budgeted for. It could be a persuasive salesman, or your own weakness when your heart overrules your head and authorises a costly impulse buy.
Readers always enjoyed writing in Bangernomics Books, so here are more facts and figures for you to ink in...

YOUR BUYING BUDGET: So the simple question is: how much can you actually afford? That figure is your budget. You really ought to know how to arrive at this figure, but just in case you can't cope: add up all your weekly/monthly outgoings and expenditures and set that against your income.
INCOME £__________(Salary/Wage/Bank Interest)
LESS EXPENDITURE £__________(Mortgage/Rent/Local Taxes/Electricity etc)
TOTAL £__________

What's left over is how much you have to cover insurance, any road taxes, insurance, fuel costs and maintenance of your next car. Is it enough? Do you have savings? Are you prepared to spend all or some of those on buying the car? Lots of questions but they do have to be answered.

SO MY BUYING BUDGET IS £__________ (in ink please). If you have not got the savings then...

BORROWING Don't. It's never good to borrow and I think you should avoid it at all costs. At best borrow from a family member who is going to be lenient and hopefully won't break your legs if you fall behind with repayments. Also an employer might give a soft loan to guarantee that you actually turn up for work. Otherwise just save up.

What can we do? Calculate to accumulate

If you haven't already done it, get a big bit of paper and properly work out exactly what your incomings and outgoings really are. You might already do this each week anyway, but many don't. It is depressing, but essential especially as motoring is so expensive these days and takes a huge chunk of your household budget.

4.Head Bangers: *Before you buy*

Bangernomics is not just about choosing a used car because it is cheap. Bangernomics can only be practised successfully if the Bangernomicist is prepared to consider all the options and think everything through. Preparation is the key to a stress free Bangernomics inspired purchase.

WHICH CAR? You would think that deciding what type of car you want would be easy. Not so. Lots of people leave their house having decided on a three-door hatchback, only to return from the showroom a few hours later in a stretch limousine. Salesmen can be very persuasive and it is all too easy to fall in love with some inappropriate and impractical car when buying from a car dealer or privately.

SIZE MATTERS Will it fit? How big is your drive, garage, car port, or even parking space outside your house, or place of work? There is no point buying a car that is going to be too big. Sounds obvious, but you should also take measurements.

Will they fit? Don't forget that it isn't just you who will be using the car. Dogs, children and spouses/partners must be able to fit in. Also, are you sitting comfortably? Bad back, tall, short, fat, thin, we are all different shapes, sizes and have different ailments. If you want to avoid the osteopath's table make sure the driver's seat is supportive and that the steering column adjusts and that you can see out clearly.

Can it cope? What do you want your next car to do? Tow a caravan? Then it needs a large engine and maybe four-wheel drive. Local journeys, then a small hatchback will do. Got a lifestyle and hobbies to follow up at weekend? Maybe an estate would be good. Only you know. There is nothing worse than having the wrong car for the job. An off-roader just for the urban school run is stupid, so is a cramped coupe' for a family of four. The thing is can you afford to buy and run it? Then that's your Banger.

WHICH ENGINE? You can have a great big debate about what sort of engine to choose. Petrol, diesel, hybrid, LPG (Liquid Petroleum Gas). The secret is getting the engine that suits you. It may be purely about what you can afford, what is available and the driving conditions. If you only do a small mileage then an ultra efficient engine may not be the answer. By contrast a big inefficient unit that doesn't do that many miles to the gallon may be fine. Ask friends and relatives what they think is the easiest to live with. Get all that opinion especially when it comes to running costs and reliability and make a decision based on the real world, then apply it to what you want to do.

IMAGE IS EVERYTHING AND NOTHING
Even though many would like to think we'd only buy a car for sensible, practical and financial reasons, actually what the car looks like, says about us is often a vitally important part of the decision making process. It is all about credibility and some cars have it, whilst others don't. A strong image means a car will hold its value well and be easier to resell. A low image makes it much more affordable. It is important to remember that often image has nothing to do with reliability, or ability. The choice is yours. Bangernomicists don't care about image, only the practicality of the car and its affordability.

EXTRA EXTRA
What a used car comes fitted with is extremely important, however it won't necessarily make it worth more, but it will be easier to live with and may eventually less bother to resell. The significance for Bangernomicists is that a lack of certain features can seriously down value a car and make it more difficult to sell, that's great news for those of us on the Bangernomics course that want value.

Good Extras Safety always sells, ABS brakes and airbags are standard fit. Just research NCAP ratings when in doubt, but interpreting them can be hard and they go out of date very fast as the standards change. Luxury and executive cars must have leather trim and automatic gearboxes so if they don't have them that's another reason why you will pay less. Small cars don't need much, although in the non Bangernomic buying world power steering is essential and air conditioning is becoming a must have to help a car sell. If you can live without luxuries, or ones that don't work, then potentially you can save some money.

Bad Extras Certainly don't pay more for an expensive music system, alloy wheels, or anything else which could be regarded as 'cheering up' an otherwise ordinary car like for instance body kit and spoilers. If someone other than the manufacturer has fitted engine immobilisers and alarms, they often can go wrong, or just go off randomly. Yes DIY add ons are bad. Also, special edition models are usually not worth more than the standard car so don't be misled into paying more.

Colour Matters, or rather it doesn't to the Bangernomicist. Anything dull such as brown, or dark ('doom') blue is regarded as bad by the car trade. Flat colours must be strong to survive in the used car marketplace. Metallic colours are reckoned to be the best, but again they must be strong. Briefly fashionable colours, usually bright ones have a habit of going out of fashion, so choose carefully, or pay a lot less. If you can live with a dodgy, paint finish or you are colour blind, bring on the badly painted car. Unless you plan on reselling the car fairly quickly it pays not to be bothered. If the cosmetics on a car are poor that is another reason to pay less. Provided the car hasn't been seriously damaged, (and you will find out how to spot this later) poor paint or a scruffy car should not be obstacle to getting a better value Banger. Whatever the colour is just say you don't like it, walk away and the seller might offer a few quid off to tempt you back.
The golden rule is buy the best-equipped car you can afford, but never pay extra, for extras. Be prepared to own and live with a weird colour and the wrong sorts of extras, or even extras (especially non-essential ones) that don't work if it means that you will get a better value Banger.

WHEN TO BUY: The used car market operates all year round, meaning that good cars will always sell, whatever the weather. Sometimes the availability and prices of used cars can change on a seasonal basis. So as well as having your wits, facts and figures about you, don't forget to consult the calendar and look out of the window at what's going on. The fortunes of the new and used car markets have traditionally been linked to a change of model or registration year. So when that happens this creates an artificially high demand for new cars and to help pay for them, a larger number of part exchanges flood into the car trade. This situation provides a variety of opportunities for the Banger buyer after a bargain.

So increased used car activity starts leading up to a plate/year change as the rate of part-exchanges starts to rise quite noticeably and that means more bargains for used car buyers. More used cars means lower prices. Auctions become noticeably busier as the part-exchanges that didn't move off the forecourt are given their last chance. So if you're confident enough to buy at Auction (see Bangers Under the Hammer).

Weather conditions can also have an effect. In the winter many can't be bothered to go out in the cold and look for a car unless they really have to. Then, as soon as the days get longer and spring officially, rather than actually, arrives, everyone's thoughts turn to forming a new relationship with a car. Obviously a mild winter can buoy up a usually lacklustre used car market, whilst a long hot summer gets everyone excited about motoring and sales reach record levels along with the temperature. But then again summer holidays when people go away can cause certain models to sell particularly badly depending on the time of year. There is a logicality about it all, so a convertible reaches a low point just before Christmas and a 4×4 isn't a high priority during the spring and summer months. However, all it takes is a washout summer to send prices and demand down. If you can anticipate and take advantage of the seasonal swings and roundabouts, good. If you need a car tomorrow, don't worry too much about the weather and just get on with it. The Bangernomics Year Planner is a bit of a distraction, but it is a pretty accurate reflection of what happens subject to registration plate and model year changes in your part of the world.

WHICH BANGER? There are thousands of Bangers for sale at any one time and the choice can be bewildering. The best approach is to draw up a shortlist of no more than four or five specific models that you think fit your requirements. As a result, you will not be distracted by other cars that might be a similar price, but are unsuitable. You may already have your own set of preferences and prejudices, but there are a few simple guidelines, which should keep you out of trouble.

PRIZE BANGER There is no photo fit for what makes the perfect banger, but if you do want to contain running costs and reduce your chances of breaking down there are some types of car you should be aiming for.

Mainstream Model: Go for a well-known make and a mainstream popular model. Every garage can cope with a service and parts availability ought to be excellent and cheap. Insurance won't be an issue either, with low insurance groups and an affordable premium on the well-known models.

Basic Specification: It is getting difficult to avoid electric windows and roofs and even air conditioning, but if you do want the minimum maintenance, hassle and expense when things go wrong it has to be the basic model. Trouble is, that will reduce your comfort levels, but then if you are simply using this Banger for a local commute you don't want anything too complicated or clever as that will cost a fortune to put right.

Engine: You definitely want one of those, it is just a case of what form it may take in the future, plus what you find underneath the bonnet depends on how you want to approach Bangernomic motoring and maintenance. If you want a high performance model or a luxury car then the bigger engine comes pretty much as standard. Ideally the engine should be appropriate for the model and how you intend to use it. However, bigger multi-cylinder petrol engines are always cheaper for the simple reason that they deliver fewer miles to the gallon. If you only cover a small mileage that might be a sacrifice worth making. Big engines are less stressed and as a result can be bought with larger mileages

Manual Gearbox: If you prefer an automatic, or only have an automatic driving licence, then there is no contest, but when they go wrong it will usually cost double the amount for a manual gearbox. A manual is also slightly more flexible, more fuel-efficient and in an emergency able to limp home on just a few operational gears.

Four Doors: Not compulsory, and indeed not an option on a coupe, but it means more flexibility and practicality for the family and business car user. Hatchbacks and estates are not only easy to live with, but can be easier to resell in some locations. However, saloons are very popular in some world markets, so the more doors the better.

WHICH FLAVOUR BANGER?

Please See Ruppert's Bangerpedia Do some research and the best place to do it (do forgive the plug) is between the pages of the Bangerpedia. However, it is important to understand the following concept.

The 7 Ages of Car They start off **Brand New**. These are showroom fresh and have delivery mileage, which can vary hugely and have nothing to do with Bangernomics, except that this is where all cars come from. A sub division of Brand New is **Pre-Registered**. Technically this is a used car because there is a name, usually a company, on the V5 registration document, which gives the car dealer an excuse to sell it cheaper. They are not being generous, just doing it to get a bonus for registering (not selling) a car from the manufacturer. **Nearly New**. These are sometimes dealer demonstration vehicles, or ones that have been on hire, or short-term lease fleets. Most cars under a year old with less than 10,000 miles can be regarded as newish. Nearly new also encompass dealer warranted used cars and what you can also find at car supermarkets. **Used**. Finally, something Bangernomic that is affordable. This is the general population for all cars being sold and includes private and dealer sales. Usually the car is warranted and usually less than ten years old and 100K miles. **Banger**. Sometimes beyond its tenth birthday and possibly, but not necessarily, high mileage. Ideally they are all depreciated out, but still full of life. Potentially the vehicle could be at the crossroads, because it is uneconomic to repair, or worn out. The next stop would be the breakers' yard. If it dodges the scrapper the Banger makes it onto the next stage. **Classic**. All of a sudden the Banger crosses a threshold. For a while it is what a pretentious person may call an **Ironic Classic**. For the purposes of amusement, a fairly ordinary and previously common model is talked about in fairly hushed tones as a future collectible. You may laugh, then the next day it's a full-blown classic car that is going up in value. A classic car is now just a very old one and the trick is to buy it before the prices start rising.

Anyway, let's look in a bit more detail at the groups of vehicles you could be considering.

TOWN CENTRE TOTS: Inner City Bangers
I'm assuming that what you want is a town centre assault vehicle and not much more. Something for local journeys and not the annual 35K mile slog up and down the motorway. What you really want here is, cheap and cheerful and above all, small. They can often be worn out, or owned by some old dear who has kept it in pristine mechanical and cosmetic condition. You just never know, which is pretty much how it go with every single used car.

HAPPY SHOPPERS: Bangerminis for Errands
Some people call them superminis, but essentially these are the second car and shopping hatches that dominate the market place especially in European and Far Eastern countries. They can be bought privately and be well looked after, when the mileage stays low and the service book full. However once they have had multiple owners their condition can deteriorate rapidly as servicing becomes optional for some careless owners. Companies use them as

pool cars sometimes and that also means that they can be in poor cosmetic condition when sold. Prices can be rather high especially in built up areas where more people want conveniently small shopping trolleys. Automatics are particularly sought after. Demand for good examples outstrips supply and mostly you will get what you pay for because there are always tatty ones on sale.

SMALL FAMILY BANGERS: This is the Golf Class that every mass market manufacturer needed to be in order to make a decent living. Hence you got brilliant rivals like the Focus and the Astra. These are referred to as hatches because that's what most of them are although the rest of world stuck by the saloon format in the mass market. Some of those made it to the UK in small numbers so that traveling reps could lock their samples away from prying eyes. These can lead a really hard lives, or have been on hire fleets. Ideally you want to deal with the careful, mature and considerate family owners who are moving onto the next one.

MINI BUSES Essentially a high rise hatchback and often called compact people carriers by cleverly repackaging the interior so there is more flexible space. However, many of these models only seat the standard five bodies, but at least have room for some luggage. Like any other vehicle they can lead hard lives and look scruffy, or be well cared for. Makes sense as family transport, but maybe a cheaper conventional estate or hatchback would do with a low Bangernomic price tag.

EX COMPANY CAR FAMILY BANGERS: These are the saloons or hatches that were basically company cars. So-called executive cars are mixed up with this lot too, posh ones with upmarket badges in some cases and lots of extras that we can buy at some very low prices. When bought by a private owner and looked after they make a great buy. Many are also bought commercially and provided you buy with few owners they still have lots of life left in them. The body might look straight, but the mechanicals could be on their last legs. Interior condition could be ragged if it has been used as a surrogate van. Prices can be low as the large numbers around means buyers are spoilt for choice.

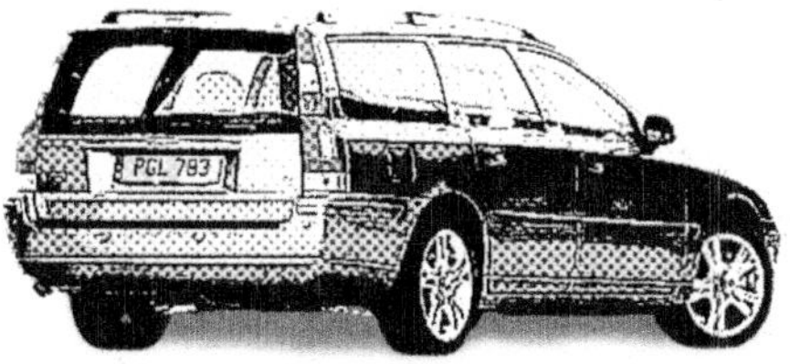

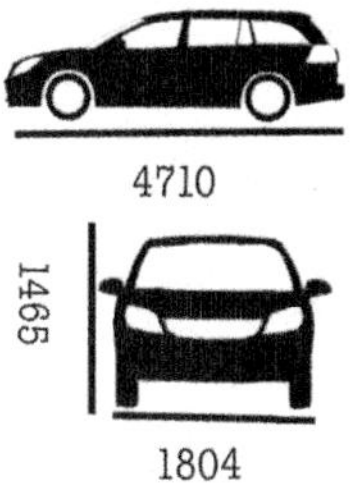

ESTATE BANGERS: These can come in small, medium and large sizes. Mostly bought to do a job and that can mean that they have lead a hard life. So whether it is coping with an extended family and their hobbies or a business user with business to do the estate can sometimes be in marginal condition. When they deputise for builders vans, the interior can be shot. Same goes for the suspension, which can take a lot of punishment. Estates can be worth more than hatchbacks and certain models are considerably more desirable than the saloons.

LUXOBANGERS Can a luxury car ever be called a banger? Well, why not? That's why we have gone a bit over the top with this one, as a Barge needs a bit more evidence to make its case as a Banger. What you get is an awful lot of car for hardly any money. For some just running a car like this for a few years is fun. Also when some pointless electrical system breaks (e.g. heated seats) there is no point in fixing it. Also big cars have big unstressed engines and components that should last longer. Our old friend depreciation also means that prices are on the floor pretty much permanently. The downside is always going to be high running costs in terms of fuel and when things go wrong. The increasing technology is a pain and can make simple diagnosis of problem an expensive undertaking. Obviously if anything breaks they can cost a lot to fix, but that's why salvage/scrap/spare parts yards were invented. You should be able to get second-hand parts at affordable prices and of course you would never go near a main agent for servicing would you?

It's all very nice to get a ton of clever extras, leather seats and a lazy automatic gearbox, but there are potential downsides. These include some fairly grim repair costs when something goes bang and with big petrol engines there will be marginal miles per gallon. However, it's no use moaning that sorting out the air suspension on your Luxobarge costs a few grand because these always were high maintenance motors. So be prepared for that and if you are mature enough to cope, then you can enjoy the benefits of someone else's brush with cripplingly depreciation. One solid theory is that costly Bangers are better engineered. Indeed some specialists have suggested that servicing can indeed be overlooked. Well, maybe do the basics, just change the oil and ensure that it is roadworthy, but provided you are not charging around the country and racking up multiple miles for business, there is no reason to be on a maintenance at all cost schedule. As you probably have no intention of selling the Luxo Banger on with a FSH (full service history), there is no reason to go bankrupt in your pursuit of Poshonomics to throw in another made up word.

Be prepared to pay a little bit more, whether you are after an executive motor, an off roader or maybe that sports car you always promised yourself. I reckon Poshonomics is as much Bangernomics as buying something basic for a few quid. It is all completely relative. You will still be buying a

BANGERNOMICS
LuxoBangers
Needs to have
Full Service History
and not be Broken
Buy Pre-
Depreciated
Avoid main dealers for
Servicing
Parts

used car at a fractional percentage of its original cost. These Bangers are cheap precisely because they are expensive to run to and fix. This means it is even more important to make sure that the Luxo Banger you are buying is in decent condition.

To be on the safe side, only buy a Luxo Banger which has a comprehensive service history. It would be nice to see some recent big bills which have been paid, so you don't have to. What you must do is brace yourself for a potential killer bill. That will either have to be paid or lead you to abandon your Luxo Banger experiment right there and then. It also pays to be creative and talk to garages, specialists and other owners. Main agents may say that they are only ones who can fix complicated Bangers, but this is not the case. Also, parts don't have to be brand new. There are lots of specialists around who refurbish and supply them with a guarantee. A salvage yard supplying secondhand parts could well be your best friend when it comes to keeping your premium banger alive. Some specialists admit that a big unstressed engine, which is lightly used, can be used until something expensive goes bang, in five years time. Here is perhaps the most comfortable way of owning a Banger and arguably the most satisfying.

SPORTIBANGERS Why not? Banger buyers can have fun too getting some style on the cheap. These days a lot of models can qualify as a performance car and the vast majority are bought by enthusiasts. The time to avoid them is when they slip into the hands of those who can't afford to run them. That is because proper servicing and insurance is always going to be higher than standard models. Performance cars with an image are more costly, but depreciation on the majority of models is quite heavy which makes them bargains, especially after they go out of manufacturer warranty. Expensive to own from maintenance to insurance, but depreciation means some real bargains.

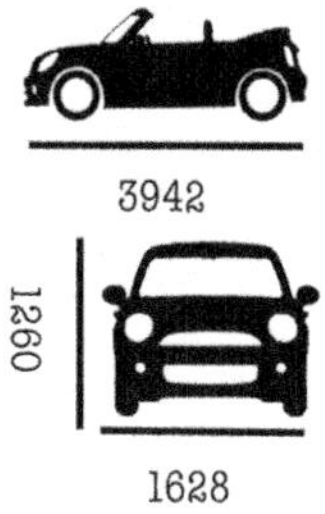

BANGERVERTIBLES Convertibles are not simply about showing off, although that is one of the many upsides. Another benefit is that you may be able to get some 4' x 2' planks of wood on board. Provided of course that you tie a handkerchief, ideally red, on any overhang. Unless it is raining, or you are parked up in a town centre multi-storey, there is no reason to ever put the roof up. That's the whole point of buying a car with half the metal missing. Just put a hat on. Mind you, those folding roofs are somewhat worrying. They really are fabulously complicated and if they go wrong, or have been mistreated, make sure it is either stuck open, or stuck closed. Then pay an appropriate amount. The simple Bangervertible fact is that car owners really don't need them, but thank goodness they do. It means that there can be a lot of low mileage, well maintained perfect condition convertibles out there. Caution is of course required because when they fall into the wrong hands you can end up with what is a skip on wheels.

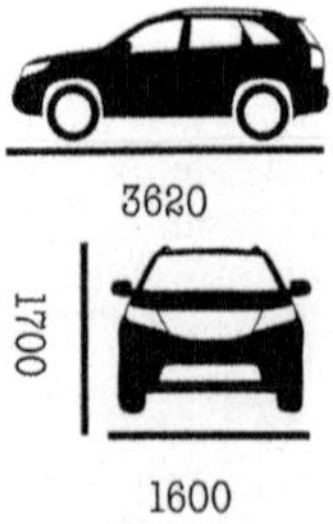

BANGER UTILITY VEHICLES Which are a bit daft if you are trying to save money but if bought to do a job can be worthwhile. Petrol engines are the death for values in Europe. They can have lots of equipment and are over engineered to the point where they could last you a lifetime especially as so few ever go off road. There are several types of four-wheel drive vehicles from so-called lifestyle off roaders, which are smaller, and the heavy-duty 4×4s designed for farmers. Many are bought for style and image rather than ability. However, some can lead harder lives. Ideally you should have a purpose in mind for a 4 x 4, such as towing or real off roading because running costs are more expensive than for a conventional car. Not only that, 4 x 4s are slower, less comfortable and don't handle like a car, hence the popularity of the smaller lifestyle models. Only appropriate for the Bangernomicist who is going to make some use of the car's capabilities. Otherwise the higher maintenance costs and petrol consumption could be big penalties.

BANGER VANS People Carriers or MPVs, multi purpose vehicles) once derided a vans with windows, these are actually high-rise estate cars. You can mostly get six or seven on board in comfort, but the drawback is there isn't room for much luggage. Being a multi purpose vehicle though means that the seats can be moved around, or taken out completely to offer more flexibility. Mostly family owned, but also bought commercially by amongst others, taxi companies. Often people carriers can be in well-used

Condition so buyers have to be careful. If you need the space then the People Carrier is perfect, but be aware of its limitations.

COMMERCIAL BANGER VANS Bought by companies and sometimes driven by psychopaths. Abuse multiplies down the chain of ownership until it looks and drives like a shed. If bought direct from a company with a full history it may make sense, but more often Mr Local Builder is unloading a rusty, mechanically worn-out heap. Rear load area and doors often rusty and damaged, suspension and the engine may also need attention. Can be attractively low priced but beware the extra tax (VAT) as this may be classed as a commercial vehicle, which needs to be paid in some cases, especially if you are buying from a company. Often run into the ground before sale so be careful. Just the two seats. Criminals like breaking in to see if there are any tools/valuables in the back. Not ideal for Banger buyers in search of a trouble free life.

CLASSIC BANGERS: The great news is that you don't have to operate at the top of the vintage and classic car market, where they are not so much vehicles as blue chip investments. The classics you should be choosing are affordable and practical Bangers that are full of character, and if looked after, resistant to depreciation. I would like to suggest that they are Classic Bangernomics, or Banger Classics. In order to bag yourself a usable classic it is best to **Pick an Icon.** There is no point in choosing some obscure sports car that will be impossible to find parts for, difficult to own and most important of all, has a limited following. The key to buying a depreciation proof classic, is finding a car that a lot of enthusiasts love. These are the iconic cars that are easy to find and have great support from clubs and specialist suppliers. We don't usually mention specific models as the upcoming Bangerpedias will help you make a choice, or you can simply email the Bangernomics office, and tell you what we think. For example, we would point you in the direction of the Morris Minor, MGB sports car and Series Land Rover. Nothing very complicated or large, you need to be able to park your classic in a garage if you want to look after it properly. **Enjoy Driving it.** When you have a shortlist the next thing to do is find out if you really like the classic, rather than the idea of a classic car. This can only be done by seeking out cars and owners. There is plenty of research to be done and no substitute for reading as many books, magazines and Internet articles as you can. It is also vital to touch metal. This means following up advertisements and looking at Bangers for sale. Ideally you should also join the owners club, attend meetings and shows. They can give you expert advice and help when you decide to buy. Most important of all drive the classics. It is not uncommon to have unrealistic expectations and actually find that your dream Classic Banger is a nightmare to drive and live with. So make sure the classic fits your lifestyle and your garage. If you can't borrow a Banger you could hire one.

BANGERNOMICS
Classic Banger
Choose a Classic Icon
with Good parts availability
NEVER BUY A RESTORATION PROJECT
Fits in your garage or under a car port
Must make you feel happy everytime you drive it

This is very important: never buy a restoration project. It may start out cheap, but it will prove expensive longer term and you may never finish it. Whatever someone quotes as the cost of making it nice again, will inevitably double, triple, or far worse.

Always get an expert to check the classic, that way you won't buy a dud, or pay over the odds. The great thing about an older classic is that it should be straightforward to service and that is something you could learn to do. Actually using the classic makes sense because it will be far more reliable, don't worry about adding miles. 'Experts' might tell you that it will devalue the Classic, however unless the Banger is very rare, or low mileage, it is the condition which will always be the important factor if you eventually resell it.

Look After it. If buying a Classic Banger is simple enough, it is vital that you take good care of it. A secure, dry location has already been mentioned and indeed finding a decent local servicing garage is equally important. Maintaining the Classic Banger will not only stop it breaking down, but also maintain its value. However there is no point taking a Morris Minor to a Lexus main agent, who will just scratch their heads and looked baffled. You need an independently run garage that understands older Bangers and can actually diagnose and repair. Indeed these older Classic Bangers are much simpler to understand and work on.

Modifications could make your Banger Classic just that bit more useable. Chat to other owners and specialists for help. Make any modifications reversible, as the next owner could want originality. Rust proofing is worthwhile. A rotting body is one of the most expensive parts of any Banger to repair. You need to find the correct product to paint on the underside and inject into vulnerable sections. A Banger Classic must be comfortable to drive. No point being uncomfortable is there? So changing seats is the very least you could do. Just stick the originals in the garage or loft. Standard old-fashioned lights are usually pretty poor by modern standards, so up-rate them. Go for modern, good quality tyres and also consider getting a set of winter tyres fitted to some spare wheels, that will get you through the snow.

Most of all though, own a Classic Banger because you want to. There is no point driving something that you don't like purely to make money when it sells or save some tax. Owning a Classic Banger is a commitment and a lot less boring than a dull old used car.

Values: The biggest and best reason to get into Bangernomics with a Classic twist is that the Classic Banger should at the very least, hold its value. Is there anything more depressing than depreciation? It's actually the most expensive word in the English language as the new car you bought a few minutes ago for £15,000, is now worth £10,000 and it will keep on falling until you have to pay someone to take it away. Just imagine then a situation where the car you own won't just hold it's value, but is more than likely to increase. The truth is that buying a Classic Banger can be good for you and your wallet. If you like cars, you'll love every motoring minute.

After all that waffle you should now be able to put together your short or wish list...

BANGER SHORTLIST

1.______________________________
2.______________________________
3.______________________________
4.______________________________
5.______________________________

BANGER VALUES Now find out if you can afford any of the cars on your banger shortlist. It is simply a case of looking through car adverts and jotting down prices of the cars you like the sound of. Ideally look at cars local to you as this will give you a feel for the market. Also if you want to buy a cheap car there is no point wandering too far for a Banger so try and keep your search within reasonable travelling distance.

If possible, understand why one car is priced higher than another, find out whether it has had one owner or thirty two, has a valid roadworthiness certificate, whether it is a private, or dealer sale, all these things make a difference.

Price guides are just that, guides either in magazine or online form. The condition of the car itself and the deal you can negotiate is always the deciding factor. Also, some price guides may only cover cars up to a certain age, so it may not cover the Banger you had mind. However, you should now be able to put an average price on your short/wish list.

BANGER SHORTLIST (compile your shortlist with rough price guide)

1.______________________________ £__________
2.______________________________ £__________
3.______________________________ £__________
4.______________________________ £__________
5.______________________________ £__________

INSURANCE

So you have sort have decided what model and maybe what make of car you'd like to buy. How about insurance then? Sorting it out now is the sensible thing to do because it could help you change your mind about the sort of car you intend to look for. It is pointless buying a car and then finding that it is in a high insurance group and you can't afford it. Shopping for insurance is easy, you just have to invest a bit of time. You just need to get plenty of quotes. So here is the briefest guide to car insurance, which uses the UK system as basis, but the same basic principles apply worldwide.

Decide on the cover you want In the UK we have Third Party Fire & Theft which covers the damage you do to someone else and their property, also allowing for the car to be stolen, or set alight. This is worth considering for an old and very cheap car, but oddly can be the same price as comprehensive and may not be worth considering. Obviously Comprehensive pays for most of the damage resulting from an accident, much better for a more valuable car and for peace of mind as hire car charges and tow in fees may be included.

Factors affecting insurance 1. ***The Car*** Ease of repair, cost of repair, performance, cost when new. Also any modifications may affect the amount the policy costs, such as tuning the engine, or fitting alloy wheels. Installing extra security measures though should help reduce the premium paid. 2. ***Your Home*** If you live in a town or city there are higher chances of theft, or an accident. Garaging a car often helps reduce the premium. 3. ***Use*** Lots of named drivers raises price of premium, also if the car is primarily used for business rather than private journeys that can raise what you pay. 4. ***You.*** You could be a bad risk because of your driving record. Usually anyone under 25 years old automatically pays a higher premium. Over 50? Good news, you are less of a risk and should pay less.

Policy Details Excess is a voluntary and also compulsory amount that you will pay an agreed sum in the event of a claim. The higher the excess, the lower the premium. **No Claims Discount** So building up a claims free driving record means the cost of the policy will be reduced each year until the discount reaches 60%. This bonus can also be protected for an extra amount, in the event of making a claim. **Special Policies** These are available for certain kinds of drivers in particular professions (i.e. from hairdressers, to nurses), or those over a certain age (mature drivers). Under this category also comes classic car policies (mostly for cars over 15 years old), where account is taken of how cherished the car is and that the mileage is likely to be restricted annually (say 5,000 or less).

Note: Getting a quote these days simply means going online and using one of those comparison sites. Simple. Except that not all companies are on the sites, so it pays to shop around still and even find individual insurance brokers who may be able to work in a delightfully old fashioned way...

Sum Up Get at least ten quotes based on exactly the same requirements and then compare the cover and additional

benefits on offer which may include free windscreen replacement, or even membership of a rescue organisation. Getting quotes online or by some remote device has never be easier. Write down the best quote you have. It will be in your phone, but you could just write down the best quote.
£_______________COMPANY_______________

OVERALL CAR EXPENSES - WEEKLY/MONTHLY________
Now let's see if you can afford the car. Try these calculations out on your Banger shortlist and see if there is a noticeable cost difference.
INSURANCE: (maths for idiots, sorry) simply divide the quote by 12 or 52 for the weekly or monthly figure.
PETROL: Estimate how many miles you will drive each year (the national average is reckoned to be 12,000 in the UK, but if you are only going to do local trips then it could be much lower, say 3-6,000). Then research the average fuel consumption figure for your Banger. Divide that figure by the mileage, then multiply that by whatever the horrendous cost of petrol may be £______ to arrive at a figure £_____ that you can then divide by 12 or 52 to arrive at the weekly or monthly figure.

ROAD TAX: Depending on where you are, there may be a charge for keeping a car or renewing the registration/number plate. In the UK it is far too complicated and datable to mention any specific rules or guidance suffice to say that you should research just what the car you think you want will cost, usually based on engine size and type, age, emissions and probably how many cuddly toys you can get in the boot.

SERVICING: Whether or not you plan on tackling the servicing yourself telephone a local garage for quotes about routine maintenance and a major service (usually expensive garage visits happen at certain set mileages) and then allow an extra sum on top for interim repairs and replacements, such as an exhaust, brakes or tyres. It is vital to find out what sundry items like this cost. If the basic parts cost a fortune then could you really afford to run it?

Basic Service £_______
Major Service £_______
Tyre £_______
Brake Pads £_______

DEPRECIATION: A tricky question as no one (despite what they claim) can really predict what a car will be worth in a year's time, but looking at the prices being asked for second-hand ones and comparing that to the price when new is the easiest way of doing this. However, when cars get very cheap they may depreciate barely at all, or even be worth the same amount if they still roadworthy and saleable.

With all those figures to hand you can now put them in a table and see if you can make a decision based purely on what it costs to run.

So now you know whether you can afford the Banger of your dreams. Remember, stick to your budgets and don't be tempted to overspend. Finally, if you are already a car owner and plan to replace it with a Banger, refer to the BYE BYE BANGER chapter for tips on how to dispose of your car. It is in your interests to sell your car first, or at least plan the sale so that the two transactions take place roughly at the same time. You will be in a better position to negotiate with a wad of money in your pocket. If you are buying from a dealer then there is always the possibility of part-exchanging your car. This saves you the hassle of selling, but you will pay a higher price because you won't get the full value for your old car. However in the BANGER BANTER chapter we will tell you how best to tackle a part-exchange deal.

Head Bangers - What Can You do...

Talk cars If this chapter has confused or frightened you in any way then maybe you don't know much about cars. Don't worry there's lots of magazines, websites TV programmes, or whatever media delivery system is preferred at the time you are reading this, which can help you out. Also speak to friends and family who own cars and ask them what they think, everyone has an opinion about cars. Most of them are misguided, mad and just plain wrong, but don't let that stop you listening. **Drive cars** Go to car dealers, showrooms, answer adverts, just go and look at, touch and most importantly drive them. Within seconds you can probably tell whether you are happy or not behind the wheel. This is absolutely vital. Also important is talking to everyone who is crucial to the decision making process whether it is your dog, or your children. Do they fit, are they comfortable and most importantly, are they happy? **Think cars** At this stage of Banger buying it is vital to keep your mind wide open. Indeed, you may start to dream bangers, which isn't a bad thing. What should happen is the right vehicle for you and your family should slowly become obvious.

5. Browsing for Bangers: *Finding your Banger*

You know you can afford it, insure it and roughly what you want. Now where is it? For the Bangernomicist, they are everywhere. The Interweb has shrunk the car market so that any vehicle can be seen in some detail within seconds. If you are buying a cheaper car, or Banger then the best advice is to keep it local. What's the point in tearing all over the country wasting petrol, or travelling costs and time looking at load of old wrecks when you can do that on your doorstep? More importantly, if you want to subsequently bring something to the sellers attention it is better to do it to someone who lives locally and also remind him or her that you are only a bus ride away. Otherwise this is the part of the Bangernomics Bible that could date the fastest, as many of these outlets may no longer exist. There will though always be professionals making a living selling cars and usually called dealers. Private sellers are unlikely to stop trying to sell their old cars, and most frightening of all criminals, scammers and fraudsters will always try and separate a car buyer from their money. Certainly most car sales have become primarily digital and that is likely to be your first contact with the car and/or seller.

INTERWEB From the comfort of your own computer terminal it is possible to surf the World Wide Web, or in reality - your nationwide web for a used car. So simply sit back and click your mouse as they used to say. There are online editions of classified advertisement magazines offering excellent search facilities, by make, model, post code and budget taking the strain out of reading every page of their paper editions. Not only that with many ads you can also click to see colour pictures, video films and probably 3D presentations. There are also countless online classified ad sites linked either to magazines, newspapers and owner's club and they are just a search engine click away. Not only that, most independent and franchised dealers now have sites so that you can look at their used stock and many manufacturers offer direct links to their dealers who operate official used car schemes. **For**: Instantaneous. Mostly up to date. **Against**: You can't test drive over the net (wonderfully old fashioned term that).

CAR SUPERMARKETS More new cars than the market knew what to do with has resulted into the over spill being stocked at so-called Car Supermarkets where hundreds of cars are on sale. Their philosophy of park them deep and sell them cheap has largely worked, although it is not always possible to haggle a better deal. Also their part exchange offers can be disappointing. Best for nearly new cars, as older stock direct from auctions can look shabby and poor value. As ever the golden rule is shop around. **For**: Huge choice. Some low prices. **Against**: Poor part exchange deals, can't haggle.

VIRTUAL CAR DEALERS As above and you can view lots of cars online, but you don't test drive. You just pay online, the cars are usually inspected and warranted, which is good. However, they don't always want your old car, which can be inconvenient for some. They usually undercut a traditional dealer. **For**: Keen prices, easy to buy, deliver to your door within the price. **Against**: Can't test drive, or haggle.

AUCTIONS Things happen fast at an auction, remember that and you won't go far wrong. If you miss bidding for one car, another one will be along literally in a minute. It pays to be cautious and do plenty of homework. That means several visits without your wallet to see how it all operates. Listening carefully to what the auctioneer says before each lot is vital. A one-owner example with a warranted mileage and full service history is obviously the sort of Banger to bid for. For the less experienced it is easy to get caught out because an auction can be a clearing-house for unwanted part exchanges and 'problem' cars. It is also easy to get carried away and bid too much. You will not be able to drive the car before sale, although with a warranted car the right of rejection exists for a major fault discovered within a limited period after the sale. There are though individual traders who will buy cars for you at auction. They charge a flat percentage fee and for the nervous and will only buy the model and specification you want. Obviously you can bid in person and increasingly online in real time and actually see the car being driven on your TV/Computer screen. **For**: Trade prices, specialist traders who will bid for you. **Against**: Very risky. **See Further**: **Bangers Under the Hammer**.

ONLINE AUCTIONS Like a traditional real life auction but virtual and in many ways even scarier. The best known at the time of writing is Ebay but here are others which are better organised and aimed at the more well off buyer after a classic or rare car. This is where a seller uploads details and pictures of the car they want to sell and then you are expected to bid on it and actually buy if your bid is the highest. Although if the car is ultimately not as described you should be able to get out of completing the transaction, there is no substitute for actually going to see and drive it. **For**: You may get a car for less. Easy to do. **Against**: You may not bother travelling to see the car before you bid. That could cause you problems, unless there are safeguards in place and maybe a warranty or guarantee system.

Bangernomic Top Tips for buying online

1. Try to bid on a car in your area that ideally you can test drive.

2. If there is a long, detailed description, listing good and bad points, and lots of photographs, then those are all good signs that you are dealing with a conscientious and hopefully honest seller.

3. So the opposite is true, avoid short adverts with few details, or ads with small and fuzzy pictures. You really can't tell the condition of a car from a thumbnail sized picture.

4. Don't get carried away, you should always check out prices for similar vehicles in prices and in other adverts to make sure you're not bidding too much.

5. Remember that in the UK at least, if the vehicle doesn't match the description, then the seller has misrepresented it, and you don't have to buy it if you win the auction. However, if you do win, and the advert is accurate, you're legally obliged to complete the transaction.

6. Never pay until you've seen the car is the most obvious piece of advice but so few people follow it.

7. Check out the sellers feedback on the advert. If they've sold a lot of cars in the past, be cautious. They could be a dealer, or someone behaving like one and shirking their tax paying responsibilities, as there will be minimal legal comeback and no warranty.

NEW CAR SHOWROOMS: Yes really. You could trace your chosen Bangers back one stage to the car showroom by approaching the sales manager and asking if they have any part exchange vehicles they want to shift. The advantage to them is that the car is sold more quickly, without paying an auctioneer's commission and they may get a better price from you than from a car trader. For your part of the bargain you need to make a decision quickly, make the payment in cash and undertake to buy 'as seen' with no comebacks. Not for those who lack confidence then. However, if you think you can handle the pressure, contact the sales manager, explain who you are and most importantly the amount you want to spend. If you are not too fussy about the type of car/Banger then you will get a faster and more positive response. If scrap car values are very low it has become possible to buy a part exchanged car for a nominal sum. The dealer does not want the hassle or the expense of paying for an otherwise unsaleable car to go for scrap. Better to sell it to you for a token amount. **For:** A used car at a trade price. **Against:** You have to act like a professional and make an instant decision.

BROKERS Although strongly associated with the new car market, independent brokers can also source nearly new, ex management and ex dealer demonstrator stock at significant savings. It is worth asking lots of questions, some cars could be those returned under no quibble car warranty schemes and could be faulty. Beware any brokers who ask for a big deposit and establish where the car is coming from before handing over any money. **For**: They do the haggling for you. **Against:** Some may string you along without having access to the car you really want. Cars on offer are not likely to be bangers.

FRIENDS & FAMILY: There are strong reasons for not buying from your nearest and dearest. What if they sell you a rubbish used car? What if they ask too much? That is the risky element in all Bangernomics transactions. However, you have nothing to lose by putting the word about that you are in the market for a car. They might know someone, who knows someone else's friend's cousin's mate with the perfect car at a knock down price. At least you stand a chance of getting a truthful report on the condition and history of the car. Well possibly. **For:** Hopefully get a car at a knock down price because you are a friend/relative. **Against:** Blood is thicker than water, so limited come back if there is a problem, which could cause a long-standing family rift.

BULLETIN BOARDS: In staff canteens, near the checkout at supermarkets and DIY superstores, leisure centre foyers and in the newsagent's window – yes those informative little post cards are everywhere. Amongst the three-piece suites and emergency plumbers are countless cars in various states of disrepair. This is a popular outlet for motors at the truly Banger end of the market, always worth a look. **For:** Cheap local cars. **Against:** Can also be cheap, local, rubbish cars.

MAGAZINES/NEWSPAPERS These could be yesterday's news by the time you read this, and in many ways the online editions have taken over and many have links to classified car adverts anyway. Also the localised editions of classified ad magazines may have a limited life with limited appeal. **For:** Loads of cars mainly nearby. **Against:** Just one, or maybe no pictures at all, plus you must act quickly to get the bargains.

CAR MAGAZINES: The problem with most car magazines is that almost all are published on a monthly basis. So if they do have classified ads by the time you get around to enquiring the car has long gone. If, however, you are in the market for a classic car or a particular make you are much better served as there are specialist magazines in these areas. **For:** If you are after a particular model of car. **Against:** Car may be sold by the time you call.

You vs. Seller Now you are ready to start responding to adverts, but now you need to be good at psychology. You could come across a decent, honest and helpful private seller with a perfect specimen to sell, or you could meet a rogue with a rough old car. You never can tell. The majority are honest, but some may have bought an awful car without realising it, or maybe they have lost all the service history and possibly their advertisement misdescribes the car. Worst of all they think their car is worth a fortune. So if you thought buying from the motor trade is fraught with danger a private seller can be even worse. The fact is you have fewer rights if you buy privately. The car must be as described by the seller. If a private seller lies about the condition of the car, you can in most countries take legal action for your losses, providing you can find the seller and want to go to the trouble of pursuing a claim through the courts. Obviously there is a huge amount of difference between a private seller against whom a buyer has virtually no redress and full time dealer who usually has all sorts of legal obligations to verify mileage and describe the car accurately. Discovering if you have found a closet trader is common sense and give-aways will be a slick manner, unconvincing reasons for selling and more than one car for sale when you call. Maybe the seller wants to meet somewhere other than at their home. That is the sort of seller you need to avoid.

MOT Online Checks: The rather large elephant sized game changer in the 'Checking a Banger' room has been the ability to research a vehicle's MOT history. All you need is the vehicle's registration number. That means if the number is obscured in an advertisement then perhaps the seller has something to hide. With the reg number go to the .gov MOT website and see what the history has been. What you want to see are very few existing advisories. Advisories that have been dealt with and no record of abject neglect. This makes it difficult to hide bad maintenance histories provided the registration number used is genuine.

THE QUESTIONS YOU MUST ASK

Most of us will respond to advertisements by telephone and that's a good idea. By all means email a list of questions, but primarily get on the phone at some point so that you can make a decision not just about the car, but crucially the seller. If they are hesitant, or evasive and you don't like their tone, what's the point? Call it all off and save yourself a wasted journey.

Repeat the advert - Yes really, you must confirm that the details are correct and that there are no printing errors.
Mileage? Clarify precise miles, as some ads just say 'average'.
Service records? You need to know if the miles can be backed up by real evidence.
First Registered? When was the car built which helps confirm which model it is.
MOT/Annual Inspection expiry? Some ads say 'long', or 'short' MOT.
Can the car be examined by an engineer? If the seller is reluctant, they can't be confident about their car.
Any warranties/guarantees? The seller might have an existing guarantee for a replacement engine, battery, etc.
Car tax/Number plate expiry? Also is it included in price? Most sellers, including the trade cash in the tax unless they agree otherwise.
Owned the car long? If a private seller has only had it a few weeks, maybe there is something wrong.
What's wrong with it? Be specific; ask about engine, bodywork, and interior. Many sellers give amazingly honest replies, others will be evasive.
Modified? A tuned engine or spoilers could change your opinion of the car, not to mention the insurance quote.
Previous owners? Less the better, obviously. Where did you buy it? Private buyer should be able to say where it came from, for instance a family member, dealer etc
Registration documents? A private seller should at least have a receipt and certainly a registration document as

proof of ownership. If there is no paperwork, be suspicious, or don't bother.

Is the car on finance? If so, the outstanding money on the loan must to be paid off before you buy otherwise you won't own the car, the finance company will.

Can I have the registration number and VIN (Vehicle Identification Number) of the car? This means you can get the car data checked against a vehicle database, which operates in many countries. It will reveal whether the car is stolen, an insurance write off, or not what the seller says it is. If they are not keen on giving you information maybe they have something to hide.

Why are you selling? Private seller should have a quick and convincing answer.

Can I test drive the car? No point turning up if you can't do this.

Do Think carefully about the answers given and decide whether, or not the car is worth following up. You don't have to agree to see the car right away, you can always ring the seller back later after giving it some serious thought.

Don't Agree to meet the seller halfway at a motorway service area. Maybe the car isn't theirs to sell. Always make an appointment to view the car in daylight looking at a car with a torch or by touch is not recommended.

So you think you like the sound of the answers and the sound of the car, in that case you move on with confidence to the next chapter, but not before doing your coursework.

Browsing for Bangers - What Can You do?

Go Exploring What you have to do is look at as many used cars for sale as you can and consider all the different buying options too. Never been to an auction? Go and have a look. Car dealers? Spend a weekend wandering around the car lots, car supermarkets and just phoning them up, because you will learn so much. **Get Chatting** Respond to some ads even if you don't want the car, see how far your questioning gets you and just what you can find out.

6.Bang Splutter Clang:
Banger Checks

Buying Bangers This isn't difficult, but the consequences of getting it wrong can stay with you for a while as you will have to sort it all out, or send it to the scrapyard. Ideally you will have paid a relatively modest sum, so it won't be the end of the world if there are problems. Before getting too deep into the buying process it is important to remember that you should never set your heart on one particular, make, model or colour. This is all about buying the right Banger and it won't necessarily be the car you first thought of. Buying a Banger is a strange and colourful journey and these are some of the things you will have to consider, no matter how depressing and irrelevant they may seem.

There is no excuse for buying an unroadworthy heap. Mileage is largely irrelevant, unless it is a diesel because contemporary ones can be complex and difficult to fix. Also, a tiny petrol powered, shopping Banger with a huge mileage could potentially be worn out. The overriding factor is always condition. At the very least your potential Banger must have a full MoT, start and stop, not make any nasty noises, belch smoke or drive erratically.

Make sure that is the case by looking carefully at the following...oh and take a mate with you. This is really important; however much you think you know, or actually don't know about cars, always take someone with you to look at the Banger. They must be able to tell you the truth and stop you making a mistake. When buying a Banger you should be dispassionate. Trouble is, this state of mind can be compromised if we either fall in love or miss something, like three wheels, when there should be four. A mate, or one of the family will be brutally honest, but caring enough to stop you getting it wrong.

As we are still at the preliminary stage, try and keep your search fairly local. The Inter web has shrunk the car market to a couple of websites and that potentially means Bangers that are even several continents away appear to be conveniently local and buyable. It probably is buyable, but it will be a nightmare to bring home. Instead, define a realistic search area. This means you can see more Bangers and if you do buy, then you won't have to travel for days to sort any problems or deal with any issues.

It is now time for Bangernomicist to get their hands dirty. As daunting as it may seem there is no reason to be frightened by the prospect of checking a car out. What must be established is that the vehicle is suitable and sound enough for your purposes; then you can call in the experts if you want to. So the initial once over is down to you because there is not point in paying for professional checks on every Banger you like the look of.

Don't worry, even if you are mechanically bewildered because the Bangernomics approach is straightforward enough for anyone with the minimum amount of common sense and reasonable powers of observation to make an informed and accurate decision. You will be told why you are doing something and not simply given a bizarre set of confusing rituals, which could embarrass everyone. As a first step to gaining confidence why not carry out the tests in this chapter on a friend's car? Then you will know how long it takes and how simple the tests are to perform. You might be able to tell your friend the true condition of their Banger.

The point has been made to me over the years that you are better off not looking at a car as that can compromise your legal rights if anything is wrong. Now I don't see the point of buying a Banger with your eyes shut and hoping for the best. If you are buying from a private seller the chances of getting money back or a successful prosecution are marginal. It is amazing what you can spot just by looking. Even if you reckon the car is OK, the final car checks before buying should always be left to a mechanically qualified professional for complete peace of mind.

And finally, you may not be looking at a car as we used to understand it. Maybe it hovers, or travels at the speed of light, in which case what follows is utterly

Irrelevant. Well actually the basic principles should be the same. That at the very least you bother to turn up and check what you are buying out. Plus if it makes a strange noise, looks rough and handles like a lame carthorse (a reference there to what came before the horseless carriage as it used to be called), then you will have spotted a PTD, Personal Transportation Device, that isn't worth buying.

BANGER SURVIVAL KIT (Basically everything can be done with or from a mobile phone now...)

CLOTHES: (apart from clothes, you can order them over the phone...anyway). You should dress appropriately and that means scruffily. Any clothing that might be improved by the odd oil stain would be ideal. If you have sensitive hands, or would rather not get them dirty, don't be afraid to wear gloves. These should be fairly thin - don't bother with mittens - but rubber gloves allow you to poke around more easily and you can also buy thin latex disposable ones. There is also the psychological factor that if you do resemble a tramp, then the seller really believes you when you plead poverty and offer to pay less than they are asking.

RAGS: No, these are not your clothes but a couple of very useful items that may have started life as your underwear. You can use them to kneel on when looking around the car, when checking the oil, or just to help you wipe away the grime when trying to get a better look at something.

MAGNET: Not so sure about this anymore. Very rarely of much use on modern Bangers. You can still pick them up in toy shops, or why not simply pinch one of those daft things that live on the door of your refrigerator. What a magnet does is stick to metal and drop off any replacement fibreglass panels, filler and it is also reluctant to cling to rust. Rust isn't such a big issue in more modern Bangers, but you should be prepared.

TORCH/PHONE: Helps you get a good look underneath the car, in wheel arches, even in the boot and under the bonnet. Make sure it works before you leave.

MIRROR/PHONE: Not absolutely essential, but useful in allowing you to see into awkward places.

SCREWDRIVER/METAL PEN: This is for poking around with (always ask the seller's permission first), so it need not be a brand new item and an old metal pen looks less threatening.

DIAGNOSTICS BOX: With the seller's permission why not ask if you have one, to plug in your magic interrogation 'box. Whether a seller would agree to it is another matter.

CLIPBOARD/PHONE: Guaranteed to strike fear into any dodgy seller. Even if you don't feel professional, it will at least make you look as though you know what you are doing. With that clipboard you should also have a pen and the check sheets in the BANGLOSSARY. This will take you through the tests in this chapter and help you assess the condition of the car.

FRIEND: This item is particularly important because they can give an objective view of the car for example: "what a heap, you're not buying that are you? you cannot be serious, get a grip" etc. Just as importantly they can also distract the seller (by chatting to them) so that you can examine the car unhindered. Of course it may help if the friend knows something about cars, but that is by no means essential. They are also useful as a human tape recorder to witness the exaggerations of the seller when they swear on their mother's life that the car is A1, or that they will give you some money off.

BANGERNOMICS BOOK: Yes that's right, keep this copy of any old Bangernomics up your sleeve, or in your pocket as a useful for reference before going into battle to bag your Banger.

INSURANCE: Just a final reminder that when buying privately you must have some insurance cover for the test drive. However, dealers must have cover for you to drive and if they do not, something is wrong.

THE FIRST FIVE MINUTES

EARLY: It is a good idea to get to the seller's residence or forecourt early. Identify the car and then have a good look around before attracting the seller's attention. You might be able to make an instant decision; usually that it is not what you had in mind, so make your excuses and leave. Alternatively, you might catch the seller trying to kick it into life. Also, ensure that you are there early enough in the day to see the car in the light. An amazing number of people are apparently happy to pay good money for a car they could only feel in the twilight. Don't make this fundamental mistake. Just as importantly establish that this where the seller lives. Believe it or not some sellers park outside a house and pretend they live there. It may be to impress you, but mostly it is to mislead you, especially if the car isn't theirs to sell. If you have any doubts invent an excuse to go into the house, ask the seller for a glass of water or to use their toilet.

TIME: You are the customer and, as we know, customers are always right. The car might be cheap, but that is no reason why you should be rushed around it in five minutes and then be asked whether or not you want to buy. Take your time.

CAR: The first thing to organise is that the car can be clearly seen. If it is parked on a forecourt, this often means that they are squeezed in door to door, or a private seller may simply open a garage door and invite you to squeeze around it, which is not good enough. Only when you can see the car clearly can you begin to make your first checks.

QUESTIONS: This is a good time to recap the questions you should have asked over the telephone/email, especially relating to the car's condition. In the flesh, sellers start to 'remember' all sorts of fascinating facts about the car, like the time it went for a swim in the North Sea. If the life story of the car is changing too rapidly and significantly, then perhaps you should back out before wasting too much time.

PAPERWORK: If there aren't any bits of paper then make your excuses and leave. It's a pain to get them replaced. What you need to establish is that the car is as described and that the owner really does own it. Lack of documents suggests there are problems when it comes to ownership. Maybe it has had loads, or a taxi company ran it. There may even be questions over the mileage, if it has done a million miles and the seller is saying a little old lady owned it and only did 6,000 miles to the Post Office and back over 10 years, they will have binned the original paperwork to make lying easier. Every scrap of paper that comes with a Banger is important. This may be your only way of finding out the true history of the vehicle. A Banger without any background information may be hiding all sorts of secrets. Ideally receipts for tyres, exhausts and windscreen wipers and other everyday items should ideally be clipped into a folder, but they probably won't be. Make sure that the registration number or owner's name is on the receipts, which means that they relate to the Banger and are not random bills from other vehicles. Never trust photocopies. Read the story of the Banger and you might find out that it has been substantially repaired after an accident, or is two joined in the middle with glue. Again, the ideal is that the Banger's service history should be in the original manufacturer's book. The stamps should not look fresh and as though they were done at the same time. They will tell you which garage did the work. Main agent servicing for the warranty period is the norm and then ideally you might see independent garage stamps in the book after three or four years. If you can be bothered, contact the garages to see if they can recall the vehicle and any issues. Be polite, they could help.

MoT Certificate. For legal reasons in the UK, Bangers over three years old need an MoT certificate unless (under current UK legislation) it was registered 40 years ago. This guarantees that the Banger is roadworthy, but only at the time of the test, because five minutes later a bulb could blow. That is worth remembering. The certificate is printed with a signature and there should be no hand written alterations. The mileage is recorded on the certificate. The great thing is that you can now go online to verify that the MoT certificate is genuine and you can see the recent MoT history online at a .gov website. You will see whether there has been a recurring problem and what the advisories are. Defects found during the MoT are categorised as either: dangerous, major, or minor. What does that actually mean? Well if something, say serious rust is regarded as Dangerous then it is a direct risk to road safety, or has a serious impact on the environment, so the engine emission settings are incorrect. The vehicle will fail the MoT and it needs to be repaired right now. Major is another situation which affects the vehicle's safety and putting other road users at risk and that includes having an impact on the environment. That results in an MoT failure and this should be fixed right away. A Minor would not affect safety or the environment, would not lead to a fail, but it should be sorted out as soon as possible. An Advisory highlights some issue, say the tread on tyres, which could be more serious in the future. Eventually it will need to be fixed. Pass, is a pass. The MoT in Northern Island is different.

REGISTRATION DOCUMENT: In many countries you can't get the annual road tax, or get a new registration plate without a slip of paper indicating ownership. So if the seller does not have this legal document, be suspicious. The car could be stolen, the document could be with a finance company because they actually own the car. Don't ever buy a used car without getting the genuine document. So make sure that it is. Look at it very closely. Are there any spelling mistakes or alterations? If so, it may be a forgery. Assuming it is still a physical document, hold it up to the light as legitimate documents usually have watermarks. If it is a credit card then a hologram will probably be built in. If checking online, make sure you are on the official government run website. This document establishes the identity of the owner in front of you, or if you are dealing with a dealer, the previous owner. However, the person named as the registered keeper is not necessarily the legal owner.

If the car is being sold privately does the name and address on the document correspond with the person selling it? Ask for proof of the seller's identity and address such as a driving licence, passport, or electricity bill. Check that the same name and address is given on the document. When buying from a dealer, record the previous owners name and address, you may want to make contact with them later. Note the number of previous owners and the date of the last sale. Perhaps this is a troublesome car being sold in quick succession? As mentioned previously, the fewer the number of owners the more likely it has been looked after.

Write down the date of registration, colour, model description (GTi, GL, etc) engine and chassis or VIN (vehicle identification numbers) so that you can check it against the car itself. Check the registration document against the car so look at the vehicle registration mark (the number plate) and VIN - this can be found on a metal plate in the engine compartment usually where the bonnet closes at the front, and stamped into the bodywork under the bonnet. Some cars also have the VIN etched on their windows, lamps, or mounted behind the windscreen on the top of the dashboard. The engine number is usually stamped on a prominent part of the engine. These numbers on the car should be the same as those on the registration document. Even if they match, have the numbers been tampered with? Areas of glass may have been scratched off the windows, or stickers may cover up an etching, which has been altered. This is usually done to hide the identity of a stolen car. If in doubt ask the seller for their insurance documents. If the car is stolen it is unlikely that they will have any insurance documents, although even these can be forged.

BILL OF SALE/ORIGINAL RECEIPT: Should the registration document be inconclusive, perhaps the seller has proof that the car was purchased and has the receipt to prove it. Maybe you can trace this back to the dealer or person that sold it. If the details are very different from the registration document maybe you should forget this vehicle.

SERVICE HISTORY/BILLS: These are reassuring slips of paper, which could prove that the car has been looked after and could back the seller's claim that they recently spent 'a fortune'. What you want to see is a folder, or file full of history rather than some apologetic slips of paper or just a stamped up service book which can be forged anyway. Read through them carefully because you must establish that they relate to the same vehicle. Do the stamps in a service book look like they were applied at the same time? If not they may be forged. If there are garage details take them down and then contact the garage and ask them about the car. Another good reason for looking through them is that there may be a receipt that says the car was completely rebuilt after being damaged or was repaired after a flood. So **READ THEM.**

ROAD FUND LICENCE/REGISTRATION PLATE: Check that it is valid, it must be if you are going to test-drive the car and also confirm that is genuine and not a copy. **Road Tax.** By the time you read this who knows what it will be and how it could be calculated. For all we know your star sign could be a factor. As a rough guide it has something to do with the age of the vehicle. It will also relate in some way to the emissions from the exhaust, so that will be engine type and size. Another factor could well be the original purchase price. In most cases we are legally required to pay the Her Majesty's Revenue and Custom office, or the authorities wherever you live for the right to travel on the road. There is no way of avoiding it. So the first and important step is to find out when your Banger was first registered, what the g/km emissions are and the size and type of engine. For UK readers, this information will be on the V5 registration and V11 renewal document.

That is the boring bit where you may have to narrow down what type of engine and year of Banger you can afford.

Either that, or it is something to jot down once you find a Banger you like the look of and can afford.

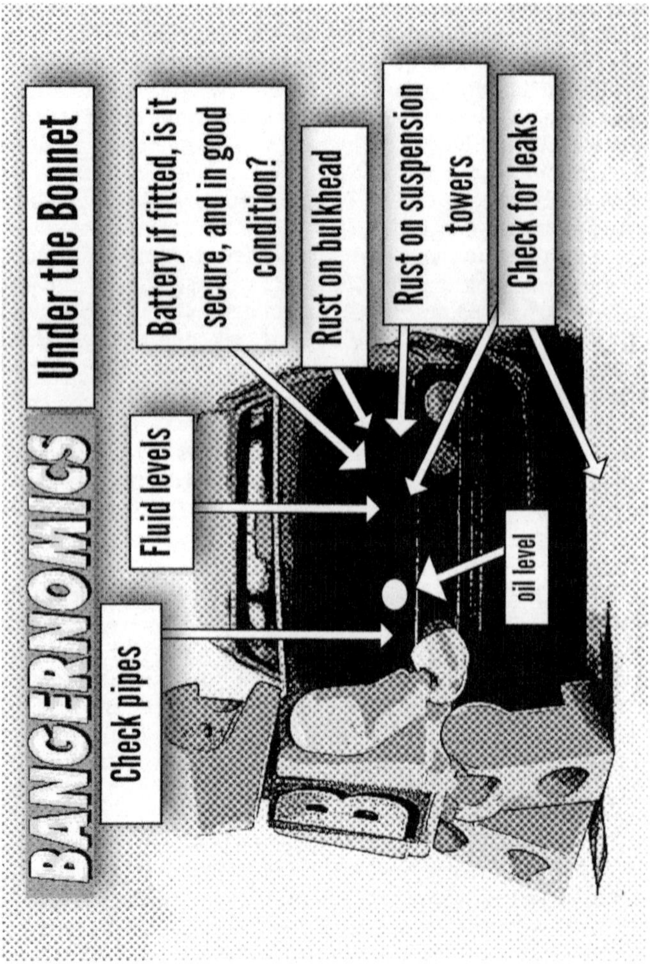

Under the Bonnet. You could look for rust and you might find some. It is bad news if you see a lot at the back of the engine bay below the windscreen, which is called the bulkhead. Also look around the suspension mounts, the two round towers above each wheel. If there is a lot of rot there, then you should forget it. As for the engine itself, when stone cold there are some simple checks. Remove the dipstick (usually coloured yellow) and look at the oil. Ideally it should be clean (honey, not black which suggests it has not be changed recently), at the correct MAX level or slightly below with no hint of bubbles or water bubbles (which suggests that the head gasket could have failed). Unscrew the oil filler cap (again yellow), which should be free of any milky deposits that again suggests a blown head gasket. Unscrew the radiator cap; can you see any oil or rust in the water? Look at the clutch and brake fluid levels too. There are lots of pipes under there, so check if the hoses are secure, perished or covered in gaffer tape, because that would be bad. Ideally you want to see a clean, dry leak free environment. The problem is that many modern Banger engines are enclosed in covers, so it is very hard to see if there are any leaks or other unpleasantries.

If you are checking the paperwork, you might as well look at the Banger itself. Start the Engine. First of all it should not be pre-warmed. Make sure the bonnet is not warm and ticking down as it cools off. The seller will have done this for a reason, usually because it's a swine to start from cold. With a diesel it might be worn out and technically lacking compression. Ask your mate to stand at the back and check whether there is loads of smoke. Play with the buttons. As you wait for the engine to warm up and adjust yourself for comfort, prod and press all the buttons to find out what does and doesn't work. Drive the Banger. Oh yes, go for a proper drive. Not simply a five minute spin around the block, but a longer revealing excursion ideally in conditions where you normally use it. In town, on the motorway, off road, just clear it with the buyer. If they want to sell it and have confidence in their Banger, they should agree.

THE CAR You've done the easy bit and if you are happy with what you have read then you can move onto the car itself. Also refer to the BANGING ABOUT chapter for more information on how a car should operate if it is properly maintained and what the signs may be if there are problems.

IN THE OPEN As mentioned earlier get the car out into the open, in daylight and ideally in dry weather. Rain does strange things to paintwork and can make it look brighter than it really is. Walk away to a distance of between 12 to 15ft (4-5 metres). Look at the car from the front, rear and both sides. Does the car sit square on the road? If not, is a tyre deflated, or is the road surface uneven? Can't find fault with the tyre or surface? Then this may indicate that the suspension is close to, or has actually collapsed. Alternatively it points to a car that may have been haphazardly bolted back together after an accident. If you feel that the car could be dangerous to drive, then don't.

KEEP YOUR DISTANCE Do the panels look even and are they damaged? By looking along the sides of the car, from both the front and rear you can see the real condition of the bodywork. Misaligned panels, ripples and general unevenness again suggests that the car has been rebuilt after an accident. You may also be able to see any differences in the paintwork colour, so have they been resprayed? Ask why.

GET CLOSER Do any of the panels look as though they need to be replaced, or simply bear the scars of careless parking? In particular look at the paintwork, especially if you noticed any contrast in the colours when standing back, inspect them more closely. Old paintwork fades and any touch-ups should be obvious. You might see some drips, flecks of fresh paint or a matt surface. A cheap paint job means that there will be evidence of overspray (paint) on the window rubbers, under the wheel arches and on the tyres. The fact that this has been done is not important, but the reason why it has been done is crucial and that means it is time to ask the seller. Has the car just been brightened up to justify a higher than average price? Perhaps some major rust or damage has been camouflaged. Ask.

OPEN AND CLOSE all the doors, boot and bonnet, do they fit snugly, or snag on the surrounding metal and do the locks work?

RUST HUNT Now you are nice and close, this is the time to start looking for rust. Improvements in bodywork protection means that it may no longer a major issue unless there have been poor repairs after an accident. With older classic cars though rust is still very much issue and it will be the difference between buying and looking for another car in much better condition. Otherwise the nasty little rust bug starts at the bottom of the car where moisture and water collects and then eats its way upwards through the bodywork. What you have to distinguish between is serious and superficial rust. The car may be full of holes but it can still be perfectly legal. Generally, minor rust on the edges of panels is rarely serious, so a rusty bonnet, doors and wings can easily be lived with, or simply repaired. However, structural rust is more serious making the car unsafe and unroadworthy as the important load bearing areas of the car can no longer take the strain. The problem is that whereas most mechanical parts can be easily and cheaply replaced serious rot is too uneconomic to repair, requiring hours of intensive rebuilding work. So if you think that the rust is serious then forget about the car.

MAGNET: This will not stick to rusted or filled areas and replacement plastic body panels. If you suspect any area, be subtle and with a magnet (put it in your palm) and cover it with a layer of paper so that it does not damage the paintwork. Be aware that some cars may have some plastic panels as standard, especially tailgates.

HANDS: These are very sensitive instruments indeed. Running the palm over the bodywork can pick up imperfections and filler very quickly. Press any suspect parts of a panel with your fingers and feel for any give, which suggests filler, or rust underneath, especially if you hear a 'cracking' sound.

UNDERSIDE Don't get underneath the car. Don't even jack it up. Even if you have axle stands. Examining the underside of a car should only be done in a workshop. However you may just be able to see some important parts by lying next to it and using your torch. Leave the complete underside inspection to the experts but you may just spot some very obvious fault.

SCREWDRIVER/METAL PEN Not to be plunged into the bodywork as the seller is likely to get upset, but it is useful to clean away mud and muck to see if there is rot underneath. The screwdriver is also excellent to tap suspect bits of metal. If you hear a pleasing ring, or solid resonate sound then may be solid enough, but a dull thud that absorbs the tap rather than bouncing it off indicates rust and possibly filler.

Bodywork. Although we don't care what the Banger actually looks like, it really is important that all the doors, open, shut and lock. Serious rot in the sills, bulkhead, or floor is not good news. Also, you don't have to put up with a shocking looking vehicle as there is such a wide choice and even cheapies can be tidy. The downside of scruffy Bangers is that they can attract the wrong sort of attention from the police, or someone who does not mind adding another parking dent.

1. SILLS: The nearest thing to you when you get on the floor are the sills. These must be solid. Also the sills can sustain damage, especially on off road vehicles that are grounded, so this can tell you something about how the car has been used and abused.

2. SUSPENSION Attached to the wheels and usually bolted to the bodywork is the suspension. You may see some leaks from shock absorbers (tubes behind wheels), which means they need replacing. Can you see any parts that seem bent or out of true, or just rotten?

3. EXHAUST There are tests you can perform with the engine running see further, but for the moment look at the silencer boxes and pipes, are they attached firmly to the car, are they rotten or crudely patched up? Some of these boxes can be replaced individually. If the exhaust looks new, or is made of stainless steel (which lasts the life of the car), you are in luck and it points to a caring owner. The catalytic converter (which cleans up the fumes) can be damaged but this is not always obvious from a visual check, again see further in the test drive section.

4. FLOOR Now look at the floor and the cross sections of metal, again you are searching for rot and signs of poor and obvious repairs. If something looks wrong then it probably is. Patches and fresh underseal all suggest a recent and possibly substandard repair.

5. PETROL TANK At the rear of the car you may be able to see the fuel tank. On older models these can rot badly (modern tanks are plastic), resulting in leaks Any suspiciously clean part would signal trouble as fuel is an excellent remover of dirt and grime. Follow the petrol line to the front of the car if you can to check leaks and condition. Obviously a strong smell of petrol indicates that there may be a leak.

6.BRAKES The thin pipes leading from the rear wheels to the front are the brake pipes. These should be intact, rust free and showing no fluid leaks. There are also flexible pipes that you may be able to see on every wheel hub (onto which the wheel is bolted).

7. OIL LEAKS Use the ground to help you. If the car is parked on its usual spot (in the street this is difficult to know, but the parking spot on the sellers property is more conclusive) then you should be able to see any oily evidence on the ground. The engine area is the obvious place to shine your torch. Also on cars that have rear wheel drive and 4×4 vehicles follow the big tube (propshaft) to the rear axle differential (T junction with the wheels at either end). If there are any shiny and oily items, then you may have spotted a leak. Minor leaks on older cars give little cause for concern, but if oil seals need replacing, this could be expensive.

8. WHEEL ARCHES As you happen to be at ground level you might as well have a peek inside where the wheels are kept. These are favourite places for mud and grunge to collect which will eventually rot the metal. Have a good poke around then and be suspicious if there is any freshly applied dirt, or possibly sticky black underseal paint that could be covering up any recent repairs.

9. TYRES You can kick them if you want, but it won't do you or the Banger any good or tell you very much unless you hear the 'ping' of metal because you missed the tyre or it is indeed, flat. Otherwise just reading the tyres could tell quite a story. It would be nice to see the same branded tyre on every axle, which suggests that somebody cared. If you don't recognise the maker, then they might be rubbish and you can do a bit of research later. Certainly they must be legal (the legal limit for minimum depth of the tread on your tyres is 1.6 millimetres, across the central ¾ of the tread around the complete circumference of the tyre). Essentially, bald bits are bad, if any tread is missing the tyre needs to be replaced and something the seller should do. Remoulded tyres are marked as such and are not very good at all, unsafe and not advisable. Also, bulges, bald spots and different patterns are bad. Badly damaged wheels with lots of chips and gouges in them point to a driver who wasn't very a good at parking. Not bad in itself, but that can mean that the tracking is out, the wheels are not balanced and would explain why the tyres are oddly worn. In itself, dreadful tyres are not a reason to walk away, but they are the four most important pieces of rubber which keep the

Banger on the road. If the previous owner has not bothered to look after them, then just possibly the Banger you are looking at has not been looked after. Tyre faults include, having incorrect pressures, cuts, lumps or bulges, exposed cord or ply and badly worn tread. A tyre that wears on the inner or outer surface indicates that the wheel is not vertical or directly in line with the one behind it. A tyre that wears in the middle or has bulges has probably been over inflated. If all that sounds a bit too complicated just remember that cuts, bald spots, bulges and mixtures of tyre types are all bad news and require sorting out. When in doubt don't risk a test drive until that tyre has been replaced. Without getting too complicated, all the information you need is written on the tyre wall itself, telling you its type, the manufacturer and maximum pressure. Personally I like to see the same make of tyre on all wheels, which indicates an owner who cared.

10. BOOT/HATCHBACK/ESTATE Open up and look inside it amazing what you may find in there. Water and moisture can collect in the nooks and crevices of the boot, so move the rubbish out of the way and look for rust. Remove the spare wheel, and look where it has been sitting, a pool of water suggests a leaky boot and probably a rotten floor section. The boot area can also show signs of repair, especially if the car has been hit hard from behind and repair sections have been welded in. The floor may also look rippled. Then check the spare wheel and tyre for condition. These can be neglected and in poor condition. Also if the battery is located here, inspect the area surrounding it, as spilled acid will start corrosion (the battery can also be under the rear seats). Is there a wheel brace and jack for changing a wheel? If you are very lucky you may find a tool kit and a warning triangle.

INTERIOR Now you can find out even more about how the car has been looked after, or not.

FLOOR/CARPETS Try and lift the carpets, in many modern cars this is not possible, but on older classics where rot is going to be an issue it is crucial to look under there. So can you see the road through the floor? A musty smell and damp carpets, accompanied with condensation on the windows, also points to lots of water inside which means rotting bodywork. With all cars look at the carpets closely. Maybe someone had an accident in there once and maybe spilt a bottle of milk. Stains and smells maybe can't be shifted so lift up the over carpets.

WINDOWS Wind them up and down (or press the buttons) do they all work? Then turn your attention to the panes themselves, if security etched with the vehicle number, do they match the number plates? If not, why not? A close look at the windscreen from both inside and out could reveal a number of scratches and cracks. If serious the windscreen will need to be replaced. Try asking the owner to replace it under their insurance policy, this is well worth a try especially if the crack is bad.

SEATS/SEATBELTS Remove any covers and assess the seats for comfort and condition. In particular, does the driver's seat adjust? Look at the seat belts, the webbing must not be frayed or broken, and the anchor points (where they attach to the body) must be secure.

MILEAGE It would be nice to know the true mileage of the car you are considering although the most important factor is always the actual condition. Far better to have a high mileage car, which has been regularly serviced than a low mile one that has only gone to the shops occasionally. Short journeys will cumulatively kill a car's engine. But if the mileage is claimed to be very low, or the seller is asking you to pay more for the car because of how well it has been looked after, then it is worth checking further. Few sellers would ever go to the trouble of renewing the interior to hide the true mileage of a car, so if the paperwork did not convince you that the mileage was genuine you can find more clues in here and spot a high miler.

• Tug on the driver's seatbelt and then let go. If it snaps back fine. If it takes forever to retract the car has done a big mileage.

• Squashy, frayed and worn driver's seat? More signs that there have been lots of bums on seats.

• A worn and shiny steering wheel, a worn and equally shiny gear knob, where the gear pattern is fading, an ignition key which has rounded edges, worn pedal rubbers and also a hole in the driver's side carpet are all signs that the car has done at least 100,000 miles.

• Look at the dashboard. Are screws holding it in place worn, or scratched? Does the steering wheel look out of alignment? If so someone may have been trying to get at the mileometer to adjust the miles. On older cars the mileometer numbers may not be level suggesting an alteration. With electronic digital readouts it is impossible to tell if it has been altered by a rogue with a lap top computer. The only confirmation of the true mileage is the paperwork and the condition of the interior.

• As mentioned before contact the previous owner whose address is on the V5 registration document and ask for the mileage when they sold it, also use the online MOT check.

• If you have an OBD scanner that may tell you if the mileage has ever been tampered with.

LIGHTS Use your friend positioned outside to check that all the lights work, front and rear from the indicators and headlamps through to stop lights and also any additional lamps such as fog or driving lights.

CONTROLS: Push, pull and prod all the relevant switches and knobs to make sure that the radio, electric windows, and sunroof all work. Electrical faults can be difficult to remedy and are very annoying. It may be something as simple as a blown fuse, then again it might not be.

Windscreen Wipers and washers: Legally they must work, even if it is not raining, don't be afraid to try them out.

BANGERNOMICS
on Bodywork + Tyres
Check Tailgate Door Bonnet Openings
Tyres must have a minimum of 1.6mm of tread across three quarters of the tyre
Damage to wheels
Rot in Sills
Rot in Floor
Damage to tyres, cuts, bulges, mix of makes?

Dashboard: Switch on the ignition and you should find a lot of lights flashing at you and/or gauges indicating something or other. Some are more important than others. **Oil** This is usually a red light, often accompanied by an oil can or dipstick symbol. Ideally it comes on with the ignition but goes out within a second of the engine starting telling you there is enough oil in the system to lubricate the engine. If the light stays on, then there is engine trouble. **Ignition**: This light may have 'IGN', or a battery symbol below it. Not surprisingly it comes on when the ignition is switched on but if it remains alight then either the battery or charging system is fault. **Brake**: A red or yellow light for both the foot and handbrake. The handbrake light will go off when disengaged. However, if the light flickers then the brake fluid may need to be topped up. Some cars have brake pad wear lights and if they need replacement the light will stay on. When in doubt consult the owner's manual if it still exists, often located in the glove box, a good seller will still have it. **ABS** many cars have anti-skid brakes and the light should come on and go off very quickly otherwise there may be a problem with the system.

Air conditioning: It may be a simple manually controlled system or a more complicated climate control set that operates automatically. Essentially it should blow hot and cold. When in doubt though it should be checked by an air conditioning specialist. It may simply need some more refrigerant gas or perhaps the sensors are not working properly. It could also be a pump or compressor which or some other part which is very expensive to sort out. This is Bangernomics though and we really should not care whether the car blows hot and cold, we just want the engine and gearbox to work.

OTHER ELECTRICS: There are a lot of extra electrics and all sorts of gadgets that have been invented and we have overlooked. Plugging the vehicle you want to buy into an OBD equipment and getting any fault codes is one option.

OILY BITS This is the part that some Bangernomicists think they might avoid, frightened by the technology. However, there is nothing scary under the bonnet and although the mechanics of the modern motor car can be complex, there is no need to know which does what, in order to spot a fault. Once again it is a simple case of keeping eyes and ears open. A few words of warning though, if you have not put your cigarette out, do it now. Also, tuck away any items of loose clothing and tie back long hair if you don't want to become an integral part of a whirling engine. Don't forget that an engine gets very hot so keep unprotected fingers away. **Note:** If the car has been warmed up before you arrive the seller may have something to hide, perhaps it is difficult to start, or it makes strange noises when cold.

SUSPENSION/STEERING: The bodywork checks, especially when you looked underneath, might already have revealed any strengths or weaknesses in the suspension. Mainly these relate to corrosion and leaks. By peering into the wheel arches you should be able to see if there is any fluid leaking from the shock absorbers and possibly corroded suspension springs.

Many engineers will tell you that mileage and age are important factors in assessing their condition. After a decade's use and/or 80-100,000 miles it is probably time that the shock absorbers were replaced. However the bounce test (which is not conclusive) does give a good indication of the suspension's effectiveness. Simply push down on the wing above each wheel, the car should bounce straight back then settle on the down stroke. If it keeps bouncing, the shock absorbers definitely need replacing.

Put some gloves on and grasp the front road wheel and push it forwards and backwards, to see if there is any play in the suspension, especially if you hear any clonks or knocks. (note: certain makes of cars have unique suspension systems, when expert help would be needed). Now rock it against the steering (pull and push) listening for clonks and watching for any play in the steering.

Put your hands on the steering wheel and move it slightly, is there any play (no more than an inch/2.5 cm), or movement before the road wheels turn? Your friend can help here. If there is play, then the steering may be worn, which the test drive should confirm. This will affect the roadworthiness of the car and may need a specialist to investigate further.

UNDER THE BONNET II: Again, don't start the engine yet, first of all try and assess the condition of the compartment. Anything that looks shiny and new is a bonus, but does it back up what the owner has already told you? If it looks dirty in there, that is the least of your worries. Of course a tidy and well-kept area would suggest a caring owner. However, dealers and owners who steam clean the engine might well end up with a surface that is clean enough to eat sandwiches off, but it can remove evidence of neglect. Steam cleaning can also cause problems with vehicle electrics and may explain why some warning lights are on.

UNPLUGGED? Hitching a car up to an electronic diagnostic system is a simple option to find out if there are any fault codes that could explain what's wrong. There are also ways of plugging a phone or PC into a car diagnostics system. If it is an option then use it.

COVERED? Many modern engine compartments have huge covers on them. This is done partly to tidy up the appearance but it also stops the untrained and unwashed amateurs like us prodding around. What you can touch is often helpfully coloured in bright yellow. Otherwise on certain models it can be difficult to see very much under there.

OIL: Look first for leaks around the engine block and gearbox. Get on your knees and look at the engine from underneath, especially the sump (a rectangular box at the base of the engine) where the oil lives. Move the car or wander back to where it was previously parked, is there an oil slick on the spot? Remove the oil filler cap and look for creamy deposits, which suggests water mixing with the oil, which is bad news.

Ensuring that the car is on level ground, take out the dipstick (a metal rod mounted on the side or middle of the engine often with a yellow lid or end), wipe it with your cloth, re-insert and withdraw again. The level should be somewhere between the max and min marks. Too low and the car has probably not been cared for. The condition of the oil on the end of the stick tells its own story. New oil is gold and clear, like honey. Normally used oil is darker but you should still be able to see the end of the stick. Dirty black oil points to a car that has rarely been serviced. The presence of tiny bubbles suggests that water is starting to mix with the oil, this usually happens when the head gasket blows. This is another problem you don't need.

Look for the oil filter; this is a brightly coloured cylinder, which screws into the side of the engine block. If it looks old and filthy, then again, basic servicing procedures have been skipped.

WATER: Now take off the radiator cap, or usually the clear plastic expansion tank which is attached to one side of the engine bay (only to be done with a cold, never a hot engine) and look at the water. Once again whitish deposits point to oil and water mixing. Put your fingers in, and then rub them together; do they feel oily? During winter especially, there should be anti-freeze in the cooling system, which turns the water blue, green or red. Orange water suggests that there is rust in the system, which in turn damages a lot of components. If you can't see any water there may be a leak in the system. Look below to see if there are any drips and look at the condition of the hoses, do they look old perished and-cracked? Orange stains are the give-away to water leaks. Again that points to poor maintenance. This area is interesting if there are any brownish stains, which may even discolour the inside of the bonnet, suggesting that the car has overheated due either to a blocked system, failing water pump, or blown head gasket. It could happen again.

BATTERY: Find it first; most are under the bonnet, some in the boot whilst the odd few can be found under the back seat.

Almost all batteries are now sealed for life, but with any other type, take off the six filler caps and check the distilled water level which should be just cover the metal elements. Check the terminals the metal posts at either end of the battery, these corrode and become covered in white powder, which can hamper its effectiveness and explain poor starting.

GEARBOX: Initially you can check the gears without the engine running and just make sure that the stick moves smoothly into all of the gears. Then start the car. The stick must not vibrate, which is a sign of wear. More simple tests can be performed on the move.

START ENGINE: At last you can start the car. Ask the seller how it is done, there may be a procedure you are not unfamiliar with or some rigmarole you have to go through to get past the immobiliser/alarm. Position your tame friend at the rear of the car, ensure the gearbox is in neutral and turn the key. Does it start easily? A slow turnover or several attempts could point to a weak battery. Do all of the warning lights go out? If something is flashing at you find out what that light refers too. Listen carefully (q.v. Nail that noise). Now get out of the car and ask your friend if they saw any white or dark blue smoke on start-up? An old car is allowed a little puff to begin with, but a constant plume of smoke indicates that there could be serious problems.

APPLY BRAKES: Push the brake pedal, it should feel firm and not immediately descend to the floor. It's essential to know they are operational before hitting the road and then maybe a wall.

ENGAGE CLUTCH: With the engine running depress the clutch, it should move smoothly and silently, a whirring noise points to a worn release bearing. Engage second gear and the handbrake, increase the revs and let the clutch out. If the car stalls with the revs dropping rapidly then it is OK.

Otherwise the clutch is slipping and may have to be replaced.

AUTOMATIC GEARBOX: You can check that the selector moves smoothly into each gear. Then with the engine running and brakes on perform the same checks, the engine should not rev when the changes are made, which should also be silent and smooth. Then, with your foot firmly on the foot brake engage D (Drive) and lightly press the accelerator, then engage R (Reverse). In both cases the car should rise up against the brakes and not stall.

DIESEL: Never drive a warmed-up diesel, always start from cold to make sure it starts easily. Difficulty starting indicates that the engine may be worn and losing compression, which is bad. Also the heavy-duty battery may be losing power and is unable to turn the engine. Be wary of an oily engine bay which could be leaking from what are known as the breather pipes, indicating severe wear. These pipes connect with the air cleaner, a box on the top of the engine. Also examine the diesel injector, pump and pipe work where it connects to the engine for leaks as it operates under very high pressure and is crucial to efficient performance and replacement is costly. Basically leaks are bad.

FOUR-WHEEL DRIVE: It is crucial to examine bodywork and underside of vehicle, especially if it is taken off-road. Look for dents, twisted chassis, suspension and corrosion. Find out how the four-wheel drive system works and then use it. Listen for noises from front and rear driveshafts and the transmission.

TURBOS: Fitted to many high performance cars and the vast majority of modern diesel engines. Best guarantee of good health is a full service history. Often it is hard to spot a turbo going on the blink, but look out for lots of blue smoke when accelerating hard.

BANGERNOMICS
Banger on the move
After the drive leave engine running, look under bonnet and underneath for leaks etc
Start from cold
Go for a proper long drive
Listen for odd noises
Prod all the buttons
If anything needs fixing, find out the cost
Pay attention to performance/handling
Don't rush into making a decision

FIRST DRIVING IMPRESSIONS: Once the car is warmed up, engage the gears and pull sharply away. If there is an audible 'clonk', on rear wheel drive cars (driven rear wheels) this could mean that the propshaft needs replacing, a major garage job. When accelerating, the car should not jump out of gear, otherwise it needs attention. An automatic gearbox should change smoothly and quietly. Compared of course, to the kind of car you are used to, does it feel sluggish? It may be that the engine needs a simple service and minor adjustment, or it could be that the engine is about to expire. Only a few more tests will tell.

CLUTCH CHECK: With a manual transmission model drive up a hill (it does not have to be very steep) and accelerate; again the car should not jump out of gear. Stop and apply the handbrake, then put your foot on the brake and release it slowly. Does it hold? Perform a hill start and see how the car reacts, especially the clutch.

SMOKE ALERT: When descending the hill leave the car in gear and take your foot off the accelerator, then provided it is safe, at the bottom, accelerate away. In the rear view mirror you may see a puff of smoke, but not a smokescreen. A smokescreen means a badly worn engine in need of an overhaul.

CATACLYSMIC: Catalytic Converters are vital but expensive components of the exhaust system and they can fail at any time. You may hear rattling in the exhaust (internal breakages) and notice that the acceleration is poor (blocked cat). Damage is often caused as a result of an incorrect air/fuel mixture, incorrect timing, or misfiring spark plugs, which could lead to a catalytic converter failure.

TAKE A BRAKE: Find a quiet road with no one about and at about 30 mph perform an emergency stop. There is no need to slam the brakes on, just apply them firmly. The car should stop in a straight line. If anything dramatic happens, like you start to go excitingly sideways, release the brake and steer your way out of trouble. End the test-drive if you feel unsafe, but bear in mind that brake problems can usually be sorted out fairly easily and quickly by a garage. Should the brakes squeal it could be a simple case of lubrication for a disc brake or, more seriously, worn wheel bearings, whilst drum brakes (fitted to the rear of older cars) could be very badly worn.

SOUND ENGINE: Now pull away with the car in second gear, or alternatively, slow right down and then accelerate in fourth. A sound engine will not jerk, stall or make any ominous knocking sounds.

STEER CLEAR: Ensuring that you are on a quiet road, bring the speed up to 30 mph and loosen your grip on the steering wheel. If the car pulls noticeably to one side there may be something wrong with the suspension, or an indication that the chassis is twisted after an accident. However, there could also be a very simple explanation such as incorrect tyre pressures.

SUSPENDED ANIMATION: Find some rough ground, or aim the car at some all too common potholes, does the vehicle pitch and wallow, taking some time to settle down? The suspension could be worn, confirming the results of the bounce test, which you should have carried out earlier. Perhaps there is a bang or scraping noise from beneath the car and the exhaust is loose or touching the road. Don't confuse these sounds with tools banging around inside the boot.

AFTER THE DRIVE: Having established that the bodywork is intact and mechanicals seem sound, or you think any repairs will be cheap to put right what you now need is a short breathing space. Getting an expert second opinion is vital. You will need time to organise this, so place a small 'holding deposit' and a simple contract. See further in the BanGlossary. Make sure that the deposit is returnable subject to car being inspected satisfactorily. Best not to talk about the amount you want to pay until the inspection is complete. If problems are found then that gives you the leverage to negotiate a better price or at least walk away from a problem car without making a huge loss and with a deposit back in your pocket. If the seller does not agree to this reasonable request then the car was probably not worth looking at in the first place.

NOTE: And if you are checking a future PTD (Personal Transportation Device) that runs on solar, steam or moral superiority power, the preceding few pages may be utterly irrelevant. However, it will still be the case that if there are leaks, strange smells, funny noises, unresponsive controls or something comes off in your hand, then you may well have a problem.

Bang Splutter Clang – What can you do?

Eyes Practice, practice and practice looking critically at all cars. Look at your own car, a friend's and get out the clipboard. Be critical. **Ears** Listen to cars that pass you in the street. Do they sound healthy? What do you think the problem may be, if any? **Nose** Just be generally nosy about cars. Drive different ones whenever you can. Take test drives whenever you can.

7. Banger Checks II: *Examining the fine details*

Although you have given the car a once over to the best of your ability there are still some checks you can carry out and experts you can use to make absolutely sure that the car is in good condition and most importantly will actually belong to you. So is it safe, is it genuine? (That explains the pretentious bona fides title then).

PREVIOUS OWNERS If you want to be thorough and perhaps have a few doubts about the car, why not speak to the previous owner? This is especially useful if the owner had very recently sold it. Perhaps there was a problem. Their name and address would be on the registration document, so you should be able to look up their telephone number. Contact them at a reasonable hour, apologise for bothering them and ask for their help. Most of us like to be helpful and most owners are happy to either brag or cry about their old cars. Simply ask why they sold it, what was the condition like and what the mileage was when they sold it?

ROADWORTHINESS TEST Some countries require this, but it is never a guarantee that a car is mechanically perfect. That is because it can easily develop a fault the day after the test and, believe it or not, some of these test documents are not entirely genuine. There are also many components that may not be checked and that can include the operation of something as important as the engine or gearbox. In most cases then you would be well advised to get the vehicle professionally checked.

GETTING IT CHECKED Just as you would probably not buy a house without first paying for a professional survey it makes sense to do the same on a used car you want to buy. For many of us buying a car is very exciting and that means we can miss many faults. Indeed few of us have the expertise to distinguish a mechanically sound car from one which could cause us problems and expense in the future. So a report on a car's condition by an engineer is vital to avoid making a costly mistake. Having an inspection also means that you can negotiate a better price as faults can be pointed out to the seller and you can make a compromise on the price or insist that seller fix any problems before you buy.

CHOOSE CAREFULLY When picking an inspection company many of the leading national companies advertise in motoring publications and on websites. It is important to make contact them and find out exactly what the costs are and what is included in the price. You also need to clarify whether the car you want to buy can be checked. Most companies aim to carry out the inspection within 72 hours, but some may be able to do that day. Not every model or type of car can be checked. Some companies have restrictions on the types of vehicle that they will inspect. They may have age limitations, for instance over ten years old so they won't look at a classic car. Commercial vehicles such as vans may be excluded and that also goes for imported vehicles. Some companies will also have lists of exotic makes that they do not want to inspect. It is important therefore to contact the company and tell them the details of the vehicle you want inspected.

WHAT IS CHECKED? Inspection companies essentially provide a condition report on the car. They obviously cannot dismantle mechanical items or be expected to predict the potential life expectancy of the majority of components. Satellite navigation systems, anti theft devices and sound systems are usually not covered by the inspection because of their complexity. These items also have little bearing on how the car works. What they can spot though is faults that affect the performance of the car or its legality and any substandard mechanical or bodywork repairs. Some companies may not inspect the interiors or report on cosmetic condition. Generally they will inspect the underside, bodywork and engine, then road test the vehicle.

COST Although some companies have a flat fee system many will charge according to the engine size. You should ask what is included in this price as there may be warranties and also vehicle status checks to see if the car is stolen, an insurance write off or still subject to a finance agreement.

AFTER THE CHECKS You should get some form of written condition report either immediately after the inspection or within a few days. In many cases you can actually talk to the engineer carrying out the inspection, although this is not always possible. Some companies may allow you to talk on the telephone to the inspector. Make sure that you clarify with the company exactly what you get in terms of access to the inspector and the details on the report and when. Once you have the report on the car you can start to make some decisions. Are there too many things to put right? Get quotes for repairs and on that basis you can decide what you are prepared to pay for the car.

ALTERNATIVELY Rather than involve an inspection company you could just talk to your local garage. If you trust them and have a good relationship with them, they should be more than happy to take a look at a car for you on their ramp, for perhaps an hours labour, or if they really like you for free. It is unlikely they would offer a guarantee, but most important of all you would get their honest, professional opinion. You could even challenge the seller, if the car passes the professional inspection you pay and buy, if it doesn't you don't and renegotiate on what needs fixing. You will soon find out how confident the seller is about their car.

UNPLUGGED AGAIN? A garage will happily hook a car up to an electronic diagnostic system but they will charge for this service. You may also be able to do this through a computer programme or a mobile phone/handheld device. If it is an option use it. The OBD readers are pretty good.

STATUS or DATA CHECK: This involves the checking of various computer databases. In some countries it may be possible to find out whether the car is registered with the police as stolen, has ever been an insurance write-off or is still the subject of a finance agreement, plus other checks that confirm/deny the car is what it is advertised as. All these situations affect the ownership of a used car so it is vital that a status check is made. Other information may also be revealed when a status check is made including the last recorded mileage and any registration plate changes. This check can be made independently of any inspection and ideally having data check first means that if it throws up any problems then there is no point having the vehicle inspected.

Banger Checks -What can you do?

Docs Familiarise yourself with all the documents that come with a car so that you will know what's missing and what may be wrong. Ask to look at friends and families vehicle documents you never know what you might spot!

Nurses Contact the companies that offer inspection services and see which one offers you the service you want and will nurse you through the procedure.

8.Banger Banter:
Doing the deal

Everything is negotiable and there is every reason why you should try to make the asking price even friendlier. Negotiation is an art and can take years to learn, but there should be enough tips here to help you.

MAKE AN OFFER Don't plough in too early with a price and delay talking about money for as long as possible. Let the seller invest time in you on the test drive and showing you around the car. Because only when you have examined the car and had it checked can you possibly say yes or no, and you should make the seller understand that this is the case. Maybe after going through whole process including the test drive and professional inspection the seller will be desperate to get rid of you at any price!

Always bear in mind who you are negotiating with. A car dealer or trader is more versed in selling techniques, but they are not all high-pressure salesmen. Many have a take it or leave it attitude, itself an effective selling technique. Generally though, stick to the facts and don't be side tracked or distracted by offers of alternative cars, or deals. Don't be trapped into committing yourself when they ask 'if I could (do this, that or the other) will you buy the car?' Just stick to your original objectives of buying the best car you can afford. The advice is similar with the private seller, who might have more experience in the psychology of selling than you imagine. Again, stick to the car and the plain facts and don't listen to sob stories. You are only interested in that car.

WHAT IS IN THE PRICE? The first thing to establish is what is included in the price. This is the time to start reminding the seller of all the promises they made when you first met. The friend you should have taken with you to stop you making a mistake can also act as a human tape recorder. So establish for instance whether any road tax is included, or whether an item such as a tyre is being replaced. With a dealer there are other issues, maybe they will service, and put a warranty on the car. With regards to the warranty read the terms carefully, ask the dealer what the cost would be without this cover and see if you can get the same or better protection elsewhere. **To sum up** if you don't like what is lumped in with the price, get it lumped out if it saves you money and it makes sense. A service would be a good thing, but make sure you get proof that this has been done before you collect the car.

WHAT TO OFFER Of course your assessment of the car and the professional checks could have revealed some faults. Here you have a choice, either ask the seller to sort it out, or make an allowance in the price and get it fixed after you buy. For a selling dealer it will normally be a cheap or easy matter for them to deal with and to be honest, do you want the hassle? Now knock a bit off the price. How much is up to you. You are aware of the market by now. Think of a reasonable offer then knock a bit more off. Don't bid stupidly low, or upset the seller. By now you should have built up some sort of relationship and if you do get along then that will make the negotiations much easier. Remember to keep your humour and wits about you at all times and the supposedly stressful business of doing a deal will seem surprisingly easy.

DOING A DEAL

Simply say, *I'll pay £__________for it.* Then stay quiet, don't justify it or hesitate, be straight and confident and then shut up. Babbling shows weakness. Ideally the seller will come back with a counter offer somewhere in the middle. You can counter again if the price seems high. Should the seller refuse to counter offer, or says no, or gives you the silent treatment then you say...*Thank you for letting me see the car, goodbye.*

If the seller sincerely wants to do a deal they will stop you disappearing down the road and offer some sort of concession. If there is still room for compromise then consider it, but never go over your original offer. That is all there is to it, be patient, firm, but reasonable and always friendly and you should get your way.

PART EXCHANGE: Remember that other car in your life? The one that now seems in relatively good condition compared to the one you want to buy. Let's assume that you do not want the bother of selling it yourself. If you are prepared to part exchange it with a dealer bear in mind that they give the lowest possible figure they can. It will not be a retail price that you could expect from a private sale, but a trade figure, which allows for any potential profit margin when it is resold. Once you understand that, then you are halfway to part exchanging your car. Usually what you want to agree then, is 'a figure to change' the net amount of money you will have to pay to get the other car.

My preferred technique is to only introduce your old car when the deal has been agreed, so that you effectively separate the two transactions. That puts the dealer on the defensive and unwilling to waste all their hard work and they may give you slightly more for your old car so that the whole transaction holds together. It is also likely that you will find out just what your car is worth.

Even before you start negotiating ideally you should contact dealers and anyone offering to buy vehicles so that you can quickly establish a rough trade price and know what the dealer is really offering. Understanding the trade value of your car is vital. Like any buyer, a dealer will respond well to a car that looks clean, tidy and well looked after, so at least give it a clean. Ideally though go to the chapter on selling your car for more detailed advice on preparation and sell privately to maximise your money.

They Buy Any Old Car Probably the easiest way out of your old car is to use a company which simply buys it for cash. This has been going on forever, it might seem all brand new and clever, but blokes have been turning up outside houses and places of works with a trailer and some grubby pound notes for generations. All that has changed is that sellers now have the options to get a bid online once the vehicle details (mileage, condition, model and year)are entered. After that it is dropped off at a depot in return for money. Trouble is that the description supplied may not match the expectation and the valuation is dropped. Be prepared for that, as they hold all the cards and are not obliged to take the vehicle. It is usually the quickest and easiest way to shift an old or not so old Banger.

CONTRACT: Once you have thrashed out the deal now is the time to get it all down on paper. The details are relatively short and simple, all you need is a brief description of the car, the date, the price, if anything has been agreed to be rectified within that amount, the seller's name, address, and signature.

DEPOSITS Although buying is essentially a case of paying and then driving (hopefully) into the sunset, there are circumstances when you may have to leave a little something as a deposit so that you can return later on. If there are a few things that the seller has agreed to sort out before collection leave a token amount of cash, keep the amount small and get a receipt. Note any sums paid and terms agreed to on the brief contract you draw up together and keep a copy. With dealers you should be able to leave an amount on a card. A credit card would give you some protection if there were a dispute.

INSURANCE Turn back to the Head Bangers chapter, I trust you did research and took notes so that you can now arrange insurance cover, Here are some final reminders and tips: Look at all the options and talk to as many brokers, specialist car clubs and motor insurance companies. Shop around, but do not choose purely on price, the cover must be suitable to your motoring needs.

PAPERWORK The registration document usually requires that both the buyer and seller sign or authorise the document and then the car can be re-registered in the buyer's name. You can find out up to date information easily and it is likely you will still be referred to as the New Keeper. You might still get to keep a bit of the old registration document until the new one arrives. It is worth remembering that usually the seller remains liable for any offences committed until the registration authorities were informed of the change. Whatever you do make sure you grab any paperwork that you are entitled to including any service history or bills. Only then can you hand over the money.

PAYMENT Traditionally the easiest way to pay is with cash. That may change and there may well may all sorts of excitingly secure ways of sending money. Ideally you want the one that costs the least as often you can be charged for sending larger amounts of money to others instantly. It sounds obvious, but only pay in full when you are actually taking the car, never, ever pay in advance. Direct bank transfers don't usually cost anything.

COLLECTION A favourite trick with less than honest sellers is to swop the half-decent mechanical parts for clapped out items. The new battery that helped to start the car first time could have been replaced by something that barely keeps its charge. Ditto the tyres - from street legal to bald penalty point items. So carefully look around for bits and pieces that look newer, or even older than before. Only then should you hand over the balance in either cash, or by banker's draft. Also remember to take a simple tool kit, consisting of some spanners and screwdrivers and possibly a jack if the car did not have one. A good idea could even be a gallon of fuel. Not surprisingly, many owners siphon off what remains in the tank, leaving you with little chance of reaching the local service station on fumes alone.

Banger Banter What Can you do?

Talk Talk Do a bit of negotiation with a friend. Ask them to pretend to sell and you pretend to buy. It could help sharpen your negotiation skills or lead to a ruddy big argument. What fun.

Double Talk Need double glazing? Didn't think so, but why not make contact with some high-pressure sellers, not necessarily cars either and start negotiating. See if you can get your whole house reglazed for free. Don't invite them around or go through with a purchase, but this is still great practice. Honest.

9.Banged to Rights:
The Legal Side

You only get what you pay for, an annoying but accurate saying. It also follows that if you buy a heap, then it was your look out and your mistake. But if someone has misled, or tricked you in some way, why shouldn't there be a comeback? This brief legal guide assumes that the seller has not disappeared into thin air, but does not allow for the fact that the legal process can move no faster than the average snail. Let's assume that you have come to an embarrassing halt within minutes, or days, of purchasing your car. After slavishly following all the advice in this book, you might be thinking about going after the author. But think again, there are several legal options; the problem is though it can be very time consuming.

Obviously the law can change, a lot, but you can reckon that the general principles here are still going to apply for some time to come. Also the consumer law varies from country to country. So here are some very general, not country specific guidelines. Essentially if you buy privately you have far fewer rights than if you buy from a dealer. Simple. Buy private, sold as seen, or buy from a dealer and unless you agree otherwise you will have some comeback.

The guiding principle should be that you are trying to buy a cheap car and in most cases you should never expect too much. If the car goes wrong then put it down to experience. If you can sort something out with the seller amicably that would be far better for everyone. If you feel you have been done over and as a point of principle want to pursue the matter then that is up to you. In my experience, small claims are never worth the bother, especially if you are up against someone who routinely flouts the law.

However, lets look at the basic principles.

BUYING FROM A DEALER

In Europe anyway you have 6 months to complain about a fault. This means that car dealers and garages have to prove if they receive a complaint within six months about a car or part they supplied that it was perfectly useable when they sold it. That's right the obligation is on them to prove it was in good condition.

In most parts of the world a car sold by a dealer must be of 'satisfactory quality', or of a standard that a reasonable person would regard as acceptable, bearing in mind the way it was described, how much it cost and any other relevant circumstances. This covers, for example, the appearance and finish of the car, its safety and its durability.

The car should also be 'free from defects', except when they were pointed out to you by the dealer, or is the sort of fault which should have been revealed by an inspection.

The car should also be 'as described' so a one-owner car must be just that. Fit for any normal purpose – so it must be a reasonably reliable car and capable of any tasks you specify to the like seating seven passengers, or towing a caravan. So dealers must describe the car accurately and must not say that the car is a good runner and if it isn't. Once you rely on a dealer's description it becomes a term of the contract, such as, 'It's definitely a 1998 model', whereas 'I think it's a 1998 model' would not be a term.

If you allow the dealer to put right the fault, however small, you have lost your right to reject the car. Alternatively you could allow the repairs, but put it in writing that you are accepting them 'Without prejudice' to your existing legal rights.

There may also be a civil action for breach of contract, if some term of the original agreement has not been met, e.g. a radio fitted, or delivery at a certain time. What you could get is a small amount of compensation. Alternatively, if the dealer is a member of a professional trade body, they may agree to abide by their conciliation and arbitration procedures. A dealer will also have to ensure that the car is roadworthy, unless it is sold specifically for parts, or breaking.

In addition to your legal rights, car dealers will be able to include a warranty. In some cases it could mean a comprehensive manufacturer backed warranty. All dealers offer some form of mechanical breakdown insurance ranging from 3 to 12 months although the terms and conditions have to be read very carefully, because there can be lots of exclusions.

BUYING PRIVATELY

It can go either way. You could come across a decent, honest and helpful private seller with a perfect specimen to sell, or you could meet a rogue with a rough old car. You never can tell. The majority are honest, but may have bought an awful car without realising it, maybe they have lost all the service history, possibly their advertisement misdescribes the car. Worst of all they think their car is worth a fortune. So if you thought buying from the motor trade is fraught with danger buying privately can often be much worse.

You have fewer rights if you buy privately. The seller must be legally entitled to sell the car, so obviously it must not be on finance, or stolen and only a data check will tell you this. The car must be as described by the seller. If a private seller lies about the condition of the car, you can sue for your losses, providing you can find the seller and want to go to the trouble of pursuing a claim through the civil courts. A private seller could get into trouble for selling a car that is not roadworthy.

Private or Trader? There is a huge amount of difference between a private seller against whom a buyer has virtually no redress and full time dealer who has all sorts of legal obligations to verify mileage and describe the car accurately. Obviously there can be a temptation for some traders to pretend to be private in order to dodge these responsibilities. The clues could be a seller with more than one car to sell and who wants to meet somewhere other than their premises. If their name is not on the registration document, then that is another big clue that they are buying cars to sell on.

Legal Action? In some cases, where only small amounts of money are concerned it certainly won't be worth your time and effort. Wherever you can, try and negotiate a settlement. In the case of a dealer most are reasonable and have a reputation to think of, so a letter to the owner, setting out your problems, politely and concisely, can produce the required results. However, if you think that the matter is serious, see a lawyer, though if the car was cheap there really is little point.

Generally there is an obligation on you to examine the car as far as it is reasonably possible, hence the phrase 'caveat emptor' which means 'buyer beware'. Although you won't be expected to dismantle the gearbox, a visual inspection and test drive would be the minimal requirements. So buying blind and then claiming the car is rubbish is not usually an option. Also, if any defects are drawn to your attention then you are stuck with any faults.

Banged to Rights What can you?

Avoid You don't really want to be involved in any legal dispute. It costs money and drags on longer than you would ever imagine. It is hassle to the power of several million, but it pays to be aware of what your rights are so wherever you live, check what your consumer rights are when you buy goods.

Decide This is your big decision, do you want some protection then buy from the trade, with a warranty. However, if you are more confident then buy private, it is as simple as that.

10.Bangers Under the Hammer: *Buying at Auction*

Here it is, perhaps the most risky way to acquire a used car. Auctions will always exist in some form or other as they are one of the few ways, (subject to the reserve price) to guarantee the sale of a used car. Virtual Auctions (which you can view on a TV or Computer screen) may take over and the same guidelines to buying online will apply. Although you can see photos and a live camera feed, you won't be able to drive the car. There is no substitute for actually seeing what you are going to buy, especially a car. So although you can click and buy, pay with a card and have the car delivered, here is the old fashioned way of buying at auction, which at least offers an interesting day out.

The point is that whether you buy privately, or from a car dealer it is unlikely that you will pay the lowest possible price. To stand a chance of doing that you need to visit an auction, but even then you could be gripped by bidding frenzy and pay more than it is worth. The other issue is things happen fast at auction, very fast indeed. If you miss one lot, another one will be along in a minute and then another and then another. Not surprisingly auctions are not for everyone. However, those who are prepared to do some research and use their common sense can buy a used car at a wholesale price. That is reason enough to consider making a bid at an auction, but before you do...

The way auctions operate is very simple. Auctions are effectively the stock market of the used car industry. Vehicles are sold on behalf of the owners, who complete a legally binding form (the Entry Form), which attests to the vehicle's age, mileage and condition. Usually a reserve value is set, which is the lowest figure the seller will accept, as the auctioneers cannot sell below this value. Often the vehicles are grouped in categories, so there may be executive and prestige sales and nearly new cars. However, it is the general sales where just about any type of older vehicle, that we Bangernomicist love, can turn up.

On the sale day itself all the vehicles should be lined up by lot order in their correct sale section ready to be sold and can be displayed under cover in well-lit viewing areas or outside. A printed or virtual sale catalogue should be available on the morning of the sale to help buyers around what looks like a massive car park. Before the sale starts there are a few hours for viewing. There may also be evening and weekend sales aimed more at the private buyer with cheaper cars although a larger number of buyers can actually increase prices.

So what happens? When the sale starts vehicles are driven into the auction hall usually in lot order. The car on sale stops in front of the rostrum and the auctioneer describes it to the buyers, a description that is legally binding. They could be guaranteeing the mileage, mentioning any major mechanical faults, the existence of a service history, any warranties and whether there is a reserve price (although they obviously won't tell you what that is). Then the auctioneer will ask for a starting bid on the car. The bidding increments are controlled by the auctioneer reflecting the interest from the auction floor. Typically bidding is done in units of 5 or 10, starting at 10 and bigger 50 or 100 increments if the auctioneer is thinks the crowd are enthusiastic. Don't worry about sneezing, or scratching your itchy nose. There are all sorts of apocryphal stories of people buying cars by mistake, but in today's professional market, that is never going to happen. Just bid clearly by raising your hand.

What cars go to Auction? Just in case you wondered, cars come from the general public, dealers, company fleets, local authorities and finance companies. Dealers can dispose of cars because it is an unsuitable part exchange as it is too old, or just the wrong make. However, a large company may simply have a policy to auction cars over a certain age and mileage. Also the car could be too expensive to put right. Quite simply the car may have been stuck on their forecourt without finding a buyer and the dealer would rather have money in the bank than a depreciating hulk in the showroom. It is rare to find ordinary members of the public putting their cars through the auction. Again, there may be something wrong with the car, a problem that won't necessarily show up on its brief drive through the auction ring. That is why you need to be so very careful. Another group of vehicles are those classed as stolen/recovered and sold by finance or insurance companies. Often they are damaged, but then they can also be in good condition. Other sources of auction fodder are fleet and company cars, which are usually no more than two, or three years old, often with a high mileage but backed up by a full service history. Vehicles can also come from large companies, or the public services such as the police and local authorities. These can lead a hard life but maintenance has usually been a priority so they may be scruffy with the remnants of a police decal peeling off the door, but still be mechanically perfect vehicle.

Your first visit must always be your research trip. Leave all your money at home and just watch what goes on. It may frighten you off and for some that might be a good thing. Auctions aren't for nervous types. Get a copy of the Conditions of Sale and Entry and any other information issued by the auctioneers. You will find out what rights you do and don't have especially as rules do vary between auction companies. Take or download a price guide with you so that you can see what the vehicles are fetching compared to the printed prices. Always remember these price guides is just that, guides and some models are more in demand than others. Also private buyers can push prices beyond what a dealer would be prepared to pay. Repair costs will be far more you compared to someone in the car trade and this is always worth remembering. Condition is vital and ultimately a car at auction and anywhere else for that matter is only worth what someone is prepared to pay on the day and that could well be far too much.

Auction Language explained

Before the sale starts listen carefully to what the auctioneer says and also read any stickers on the car windows carefully.

'Without reserve' This means that the car will be sold to the highest bidder.

'Reserve' This is the minimum the seller will want for the car, not usually revealed before sale, unless it has been entered previously and didn't sell. Sometimes the auctioneer may give a clue to the reserve.

'As seen' What you see is what you get, including all the faults. Some cars fall into this category merely because of the seller's policy.

'All good', 'on description' or 'No major mechanical defects' This sounds promising, but it only applies to the major components, So although the engine, gearbox, axles, steering, suspension and brakes may be fine, instruments, electrics, tyres and trim may be broken or damaged, Some auctions may offer an insurance backed warranty to cover this.

'Warranted all good' The auction actually guarantees that the car is as described with no major mechanical faults, so a warranted mileage and service history. A successful bidder can therefore reject a car if it is not up to these standards.

'Specified faults' A fault can be underplayed so a phrase such as 'worn gearbox' could actually mean that it is permanently stuck in second. It is worth bearing in mind that an auctioneer is acting as an agent for the seller and may only know as much about the car as they have been told.

Auction Survival Kit - what to take with you (Your phone will be able to most things)

Cash: the minimum amount required to pay a deposit although a card can be used, read the conditions of entry to find out how much you need.

Friend: to stop you making any rash bids, or buying something that you don't need. They can also drive you to

the auction so you don't have to pay a delivery charge or make a return visit.
Old clothes: after all you will be crawling around getting a closer look won't you? Better to dress down.
Torch/Phone: All the better to look into those nooks and crannies.
Price guide: Printed or downloaded (probably on your phone) price guides do what they say. Just research the sorts of cars you want to bid on and get an idea of what you want to pay. Always remember that the final 'hammer price' is not what you will pay and there will be other fees and expenses involved in getting the car you bid for home.
Tool kit/gallon of fuel to help you get the car you bid for home, but don't carry the petrol can around with you, obviously.

Before Bidding Buy, take or download a sale catalogue. The cars will be parked together and on the windscreen will be a Lot Number. This gives some indication of the running order, but don't rely on it. There may also be some information about the car such as make, model, mileage and perhaps a brief engineer's assessment, which may not be binding anyway. Watch out for 'sold as seen' and 'recorded on insurance register' (that could mean it has been written off by an insurance company as a loss). Don't pin all your hopes on one Lot. It could be withdrawn, you might change your mind, or you might not bid enough to buy it, so draw up a shortlist of Lots.

Checking the car There isn't much you can actually check. The cars are usually parked so close together that will restrict your view and mostly they will be locked and won't be started until a few minutes before they go into the auction hall.
Tyres for their condition tread and brand.
Entry form on windscreen read carefully for major faults warranties and mileage.
Lot Number so you know roughly when it will appear in the sale.

Bodywork Look for rust, poor repairs and panel alignment.
Trim anything from a cracked windscreen to missing or damaged accessories.
Interior Look for wear and tear, best done when the car is in the auction hall.
Engine. Are there oil leaks where it is parked? Be there when it is started so that you can check for smoke and listen for noises. The auction driver will usually let you look under the bonnet when it is in the auction hall.

How much to Bid? Now is the time to set your budget. Look at the price guide trade price and bear in mind what similar vehicles made at the previous sales you attended. The questions are what can you afford and more importantly what are you prepared to pay? Obviously there is no point bidding up to what a main dealer would charge for a car with a year's warranty now is there? There may well be some local or national taxes to allow for. So listen carefully to what the auctioneer says about the Lot before they invite any bids.

Auction Action a 10-point Bangernomic bidding plan
Before your lot is called position yourself near the car you want. Use your friends if the lots you are after are close together.

• Watch the car being started. Does the oil light go and off quickly? Does it start first time? Does the engine sound healthy? Are there any strange noises? Is there thick black smoke from the exhaust? When the car moves off are there any creaks, groans or scraping sounds?

• When driven to the auction hall it will join a short queue of cars. Now is your chance to take a good look inside. Maybe ask the driver to pop the bonnet. Make your decision now; do you want to bid for this car?

• If you want to bid, place yourself in view of the auctioneer and listen. What they say now is crucial. Listen out for the good points like a warranty and guaranteed mileage and also any negative ones like 'gearbox trouble'.

Now programme yourself, what is your budget? And if you have a friend with you remind them to stop you bidding above that limit.

• When the auctioneer suggests an opening bid and usually it is an optimistically high figure. Don't get sucked in too early. Wait and see what happens.

• When you bid, raise your arm clearly and confidently. Be aware of the margin of the bid: is it in 5, 10, 100, or more? If you want to make your bid less tell the auctioneer. Don't worry if the auctioneer ignores your bid because they usually concentrate only on two bids at a time.

• If you seem to be biding against someone you can't see, it is permissible for the auctioneer subject to their company rules, to take 'bids off the wall'. They do this to get closer to the reserve price. So don't get carried away and exceed your bidding budget.

• If you drop out of the bidding a shake of the head is sufficient to indicate this to the auctioneer.

• If your bid is successful you will be required to pay an immediate deposit, usually 10% or a fixed sum in cash. (as you may have noticed there are lots of Ifs at an auction).

After the sale The balance will need to be paid within a short period usually with 24 hours, or sometimes before the auction closes, by cash, or credit/debit card. You may also have to pay an indemnity fee on a sliding scale according to the cost of the vehicle which insures you against the car being stolen an insurance write off, or still on finance. Bear in mind that if you delay in paying the full amount you will probably be charged storage fees, so don't delay.

If your bid successful but it does not exceed the reserve figure set by the seller, then the auctioneers will contact them to find out if they will accept less. In the meantime don't bid for another car or you might end up with two! Should the seller make a counter offer the decision is yours, but don't exceed your budget, or pay more than you think it is actually worth.

If you have a bought a vehicle that came with a trial, in most cases you will only have one hour after payment to drive the car and find a serious fault. If a car is sold on an independent engineers report (attached to the windscreen), or 'with no major mechanical faults' then that is exactly how you should expect the car to be. Remember that it is up to you to check the car's overall exterior visible condition - paintwork, trim tyres and the interior, seats and carpets - for example - prior to sale. Never forget the phrase, 'sold as seen'.

Collection: Once you have paid the office should be able to give you all the relevant paperwork including service history and registration document, plus a sound system front, or anything else removed from the vehicle for security reasons. You will also get a pass out form allowing you to take the car off the premises. If the registration is missing, the auctioneer's receipt is sufficient proof of ownership until you can apply for a replacement.

Before you drive off though make sure that your insurer will cover you. Also remember that the vehicle has to be roadworthy. One balding tyre, dodgy wiper blade is illegal and ought to be fixed. It is not unknown for police to park outside the auction exit and catch illegal vehicles.

When you get the car home sort out any problems and as a precaution and unless there is any evidence that it has been done recently, have the car serviced.

Bangers Under the Hammer - What can you do?

Walk The best way to learn about an auction and find out of you are comfortable with the whole experience is to go to as

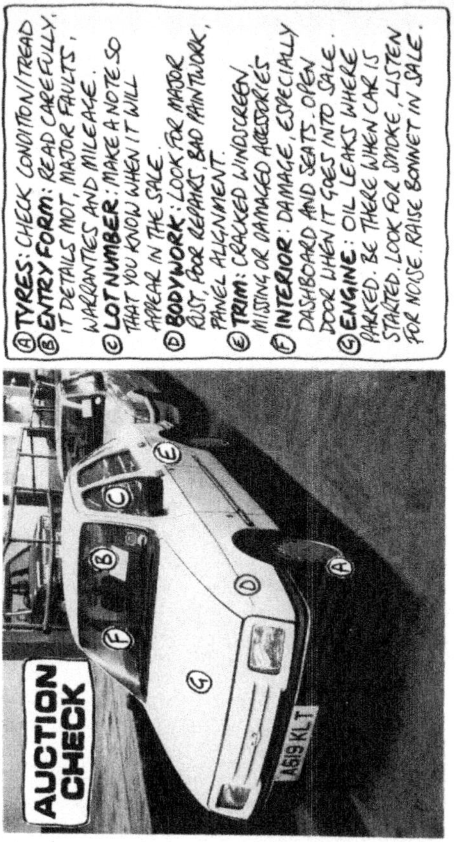

many as possible. Don't bid, don't take any money and just look and learn and walk around for several hours.

Write down prices. Compare and contrast with the prices in classified ads. They might not be too much different. Look at the condition of the cars at the auction. Would you really buy or at least bid for one?

11.Banging About: *Banger 'Care'*

You have safely steered your banger home, it is taxed, insured and possibly freshly serviced it. What next? Well you could give it a clean. Bangernomicists can now divide themselves into two categories. Leave it as you found it, or beautify and personalise. At the very least you ought to get rid of the more disgusting leftovers belonging to the previous owner, banana skins, crisp packets and old passengers that seem reluctant to leave. This can be done quickly, easily and without resorting to expensive specialist 'products'.

BODYWORK All you need is some hot water and a sponge to get rid of the grime. If you have some car shampoo use it. Rinse off the suds with some clean water and that is it. I won't go over the top with precise washing and waxing tips, but now might be a good time to check inside for water leaks, usually through broken window rubbers.

ENGINE Resist the temptation to clean under the bonnet. Water and engines don't mix, they have a habit of refusing to start, or suddenly developing intermittent electrical faults. What you can do though, is wipe the grease off the items that you will need to check periodically, Use a rag lightly dipped in petrol. If the car has not recently been serviced to your knowledge, then a simple oil change will help.

WINDOWS For your own safety, you ought to be able to see out. The traditional household window cleaners aren't really suitable for cars as they polish the glass and leave a reflective glare. The biggest problem, even after the glass has been washed, is the presence of grease, which smears all over the windscreen. The simple answer is to use the pages of a newspaper which will soak up the grease. Just rub on and see what muck comes off. Don't forget to rub the paper over the wiper blades to prevent the grease returning.

INTERIOR A good old-fashioned dustpan and brush, or vacuum cleaner, will do the job. If any of the carpets or mats can be detached from the floor, get them out. Domestic furniture polish will do if you want to buff up plastic seats and dashboards, otherwise a damp cloth will soak up the worst of the dust. If there are some persistent smells, or festering stains, don't be frightened to use well-known brands of disinfectant.

BANGER MECHANICS If it isn't broken, don't even attempt to fix it. This has to be the Bangernomics watchword when it comes to fiddling about with anything mechanical. Once you start interfering, the fine balance of the car is disturbed and things will start to go wrong, even if you know what you are doing. Some Bangernomicists may care to regard the bonnet as a sealed unit, and simply rely on the experts when trouble strikes. That is one way of approaching it, but of course prevention is always better than costly cure. So a minimum of maintenance and observation should keep everything running smoothly especially if you want to avoid coming to an embarrassing halt.

Cars have certainly become a lot more complicated and there is much less that most of us can do in a DIY capacity. The fact is that the view under the bonnet is either dauntingly complicated or it actually looks like a sealed unit because there are big, dumb plastic covers. This depressing sight is only relieved by playful yellow screw tops that we are allowed to touch. Never mind all that, here is a brief mechanical numbskull's guide to keeping the cogs turning on cars with conventional petrol and diesel engines. For anything else the general principles for looking after these new fangled PTDs (Personal Transportation Devices) is going to be similar. Paying attention to what the owner's manual says, being methodical about certain procedures and finding a local servicing agent, or expert is still important. So I'd still read this even if your transportation device hovers.

WORKSHOP MANUAL: This will save many frustrating hours searching for various items and deciding which spare part is right for the car. They can help you trace a simple fault and are invaluable for telling you how to take bits off properly. These can often be picked up for older cars second-hand. Otherwise there is always the local library. If you can't find or borrow one I would argue that however complicated cars are these days even buying a brand new manual is a worthwhile investment. You'll be amazed at how satisfying it is to repair something simple like a windscreen washer and the details in the book at least means you will be poking at the right piece of kit. Compared to what a garage may charge for sorting that problem out the book will easily pay for itself. You will have to check online for more recent models as printing has gone out of fashion.

OIL: At the very least the first thing you should do is change the oil unless you have proof (receipts) that it has been done recently. In simple terms the oil provides lubrication. When it gets old and black, with bits of metal in it, then the liquid is less effective and starts to damage the engine. So changing the oil and oil filter is a good place to start and in most cases finish. If you don't want to get your hands dirty, ask someone else to do it. The more enthusiastic can perform a simple service by referring to the manual.

JOIN THE CLUB: If there is a national, or local club for your car or marque it could well be worth taking out a membership. Often they negotiate discounts on insurance and other motoring services plus more importantly they have a huge amount of specialist knowledge. If you have a technical problem they can usually make a useful suggestion. Some also have a thriving parts service and find items that are impossible to source elsewhere.

RESCUE/RECOVERY SERVICE: Although the cost of joining might even in some cases exceed the value of your banger, it is essential to invest in this service, especially if you plan to rely on the car. The only exception might be if you restrict the car's adventures to local runs. These days it is a very competitive market and there are some very low membership costs for a basic rescue/recovery service. However some companies may charge more for older vehicles, or ones they believe to be less reliable so always shop around and look at the small print. As with any service try to compare like with like.

TOOL KIT: A motley collection of spanners and screwdrivers are all you really need. Adjustable spanners make sense, although the purist mechanic won't be impressed and there are some good budget spanners around. The only specialist tool would be a spark plug spanner. Self-locking grips, sometimes referred to as a 'Monkey Wrench' can prove useful. A collection of screwdrivers both with a flat and cross heads is a good idea. Also a good wheel brace is worth having as sometimes the pathetically small spanners supplied with the car have little or no chance of removing the wheel nuts. Other items include: fuel can; cloth, adhesive tape; spare bulbs and fuses; tyre pressure gauge; foot pump; battery charger; and, jump leads. Buy good quality items and they will last for the length of your Bangernomic career. Oh yes and stick them in an old bag, there is no point having them rattling around in your boot which will just annoy everyone. And finally put some old clothes in the bag. If there is something you can fix easily you won't want to get your decent clothing dirty. However stupid you look in an old pair of trousers and jacket they will keep the worst of the dirt away.

WARRANTY: This is essentially an insurance policy which covers your car against specific mechanical failures. Not everyone bothers but maybe they should because on some cars it is certainly worthwhile because they are so pricey to fix. You can even get a policy, which covers maintenance and effectively spreads any servicing and repair costs over the year. It will suit some drivers perfectly and remove any worries they have about unexpected expenditure. To get the best warranty that suits you it is important to read the small print to see exactly what is covered. You may have a claims limit and may even have to contribute to any costs. Indeed high mileage vehicles and older Bangers may not be eligible for cover at all.

ADVANCED TOOL KIT: If you do take DIY car maintenance much more seriously and intend to do a lot of work yourself then I'll assume you know what you need, I would like to issue a little health and safety notice at this point.

HEALTH & SAFETY MESSAGE: It is never properly comfortable underneath any car and the only truly safe way to work is using a professional two or four post hoist that you can find in a garage, or failing that an inspection pit. In both those situations the chances of being injured by the car falling on your head are very remote. Unfortunately many home mechanics are damaged because they do not take the right precautions. Bricks, timber, breeze blocks are all inappropriate ways of balancing a car in the air. Also a jack should not be the sole item keeping a car airbourne whilst you work underneath. Indeed the scissor type jacks supplied with most cars are only intended for use when quickly changing a wheel. If you intend to regularly work underneath a car you need the right equipment, this means hydraulic jack. However you must use the equipment properly because the approved jacking points usually at each corner of the sills adjacent to wheels is really only suitable for changing wheels and it is usually only a quick way of damaging the underseal. Hydraulic jacks should only be positioned on suspension cross members, or on rear wheel drive cars the differential. Never locate the jack on the engine sump or floor. Once raised then axle stands should be used to support the car. Even so many mechanics still do not feel completely safe when a car is on axle stands alone as some cars could still be pushed off by someone leaning on the car. Ramps can also make sense, but ideally you should only put to a car on them with help and the wheels need to be chocked properly (put something heavy behind wheels) to stop the car and the ramps moving. Yes there is more to getting a car up into the air than you imagined. If you have any doubts don't do it and always consult the manufacturer about the positioning of axle stands and hydraulic jacks.

There is lots of other safety stuff too, like don't smoke when you are looking into the fuel tank and don't change the oil when naked, that sort of thing. I reckon you have enough common sense to decide what's right and what's wrong. Just be sensible, don't take any unnecessary risks and you won't hurt yourself. If you don't have the skills then...

DON'T TAMPER WITH: brakes, suspension, or steering. For the enthusiastic, though incompetent, apprentice mechanic, steer clear of fiddling with parts that could endanger your health and other people's lives.

ROUTINE SERVICING: Get it serviced. Either you do it, or get someone who is qualified to do it, this is vital to keeping your banger healthy. Service intervals vary from car to car, so an inspection could occur at between 1,500 to 3,000 miles for older classics through to 12,000 for more modern cars, whilst some state of the art cars only require a 20,000 mile or more check. Everything depends on the vehicle's age and the type of engine. You will find details of what is involved at each interval in the workshop manual or owners handbook. Even if you only cover a tiny mileage annually, you should still have a service at least once a year, otherwise minor faults may develop into major problems.

YOUR LOCAL GARAGE: Potentially your best friend when it comes to keeping a car in tip top condition is more than likely to be your local garage. If you go to a main agent with plate glass and pot plants then you are going to be paying for all that. A small independent local garage will have a much lower hourly labour rate. It isn't just about money of course you need to establish that the garage is a good one. You can only find that out by asking around friends and neighbours. Personal recommendation is the best there is but I also feel that a bad local garage will soon get found out. It is in their interests to deal with their customers honestly and fairly otherwise they will run out of business. I have found that a good garage won't do any unnecessary work and will do their best to solve problems economically. When in doubt though make contact with your local trading standards office to see if there have been a series of complaints against the garage. Also membership of a garage association should mean that there are complaints procedures and certain standards required for membership. This is still not a guarantee that you will be happy with the garage. Essentially you need to get on with them and they need to understand what you want and what you can afford. A good garage will never do work without consulting you first. Spending time finding the right mechanic is just as important as finding the right Banger.

BANGERNOMICS GOOD GARAGE GUIDE As well as local recommendation, why not join a car club specific to you your car? Most have recommended garages that as well as being experts in particular may offer discounts on parts and servicing to members.

Once you have found your garage you then need to set your own agenda and be aware of the costs involved. Money seems to be at the root of a lot of service related problems. Calculating the cost of work is not difficult. There are manufacturer standard times for all procedures, whether it is unscrewing a bolt, or a major crash repair. Just call the manufacturer to find out the times and then all you need to know is the garage's labour charge. This charge obviously varies depending on their location. Multiply the figures and not only can you check their sums when you have the bill, once you have a figure you can now shop around for a better quote. Getting a full quote should alert you to the worst of the bad practices such as charging for oil by the litre and being charged under sundries for the mechanics rubber gloves.

Ensure that any local taxes and parts costs are included otherwise there will be a nasty surprise. Always remember to put what you want done in writing, state a collection time and the minimum you agree to spend. This saves a lot of confusion and expense later. Most importantly make it clear that you will authorise any additional work. It is not uncommon to be presented with a big bill and a new engine simply because you insisted on the car being ready by 5pm. When it comes to the 'can't get the parts' excuse, most garages have standard service items in stock, but the majority of parts can be delivered within 24 hours. Oh yes and once the car has been fixed you can ask to see the parts that have been replaced.

If you do have service and repair related problems, talk to the service or garage manager straight away. If the problem cannot be resolved, you will still have to pay the disputed bill to get your car back. However, give the garage a letter, detailing your complaint, saying that you are dissatisfied and that payment is made 'without prejudice' to your legal rights. This will ensure that the garage cannot say in court (if it ever gets that far) that you accepted the repair because you paid the bill.

This may come as some surprise to you, but running a Banger is just like running anything else with four wheels and an engine. Your motoring mindset just has to change slightly. There is plenty that you can do as preventative maintenance. This may be something that you have not done before. You could also make your life a whole lot easier by finding a great local garage or specialist company to do that for you. It is also best to assume that you know nothing about cars, even if you think you do. Unless you have worked in a garage and have the relevant qualifications then it is best to leave nothing to chance.

Always be on the lookout for problems and either catch them early or prioritise any issues before they become expensive problems. Here is a very basic guide to keeping an eye on your Banger. It isn't full proof and it certainly isn't fool proof. Let's take a look at some of the most potentially expensive things that may go bang.

Listen to your Banger. Once in a while, turn your radio off and tell you passengers to politely be quiet. That's because your Banger is talking to you. No really, it is. Whether you are going up the motorway or idling at the lights, your Banger could be making a sound that is a cry for help. It pays to catch any problems early. The longer it takes you to take it to a garage or a specialist, the more expensive and possibly dangerous the issue could become. Here's what you should be listening for.

Pops. Let's start with an obvious one. A pop from your exhaust may indicate that there is an issue with the fuel system. The mixture may be wrong and needs adjustment. With old Bangers that was easy, modern motors will need to be reprogrammed. Not only is it annoying it may affect the vehicle's ability to pass the MoT. The fuel consumption can also be adversely affected. Then there is issue of potential expensive damage to the catalytic converter and exhaust system.

Squeals. Often that is the brakes. They could just be badly adjusted. At the very worst the pads could have worn down to the metal studs. If left too long that would score and damage the expensive brake discs.

Screech. There are lots of belts under the bonnet, which power the steering, charge the battery and operate the air conditioning to name just a few parts. If badly adjusted or at the end of their useful life, may start to screech.

Scrapes. At its very simplest it could be worn wiper blades which are in such bad condition the metal part of the wipers could be scraping the screen. Otherwise it may be a part of the bodywork catching a moving mechanical item. That has to be dealt with quickly.

Clunking. More often than not this happens when the Banger's suspension bounces. That could be broken or serious wear to the suspension and possibly steering system.

Clicking. With the steering turned to the extreme left or right, clicks suggest that the steering joints are badly worn.

Ticking. This is usually heard from the top of the engine.

When the engine is started this should go away quickly, otherwise there is the possibility of a lack of oil in the engine. It is a simple fix, but can cause serious engine damage.

Hissing. This is another engine sound and is a prelude to the engine overheating, or that there is a problem with the exhaust system and a catalytic converter, which could be blocked.

Grinding. If you hear a grinding noise it usually comes from the transmission. When you have any resistance or just hear a metal on metal noise, that is serious. It will be even more expensive if you have an automatic gearbox. Grinding can also be traced to the wheel bearings, which eventually wear out and need to be replaced.

Whines. Whining can also be associated with worn out wheel bearings. Otherwise it could be part of the automatic gearbox system or some other mechanical part needing lubrication.

Unwell Catalytic Converter? You never need to see your cat, as a catalytic converter is often referred to, but it is one of the most important parts of your Banger. It helps reduce harmful emissions and also means that your Banger will pass the MoT. So it helps to know when the catalytic converter may be failing. That may be happening if the fuel consumption rises. It is worth getting into the habit of checking your mpg from time to time. Use the on-board computer when fitted, or could even log it every time you brim it and write your mileage down, if you are slightly odd. Otherwise monitor how often you need to fill up when using the Banger normally. Any sudden changes might suggest that there is a problem. If your Banger does not start it could be for a variety of reasons. With an old high mileage vehicle it can become clogged up and the smoke and gas from the engine finds it more difficult to escape. Acceleration may become sluggish. Again, there are a number of reasons why your Banger is not as keen to pick up speed, but an old, failing and possibly blocked catalytic converter is a good reason why this is happening. It also pays to listen to your cat and yes it is worth doing from time to time when you may hear a 'tinkling'. The metal elements inside a catalytic converter can break up and that would be the sound of them moving around. And finally, your cat may be failing if the engine light comes on. So you should stop driving immediately. Many motorists don't. So if you spot any of these symptoms it would be worth paying a visit to your local garage before your Banger fails the MoT. The garage will have the electronic testing equipment designed to report any engine malfunctions including the failure of a catalytic converter.

Blowing head gasket? The head gasket is a vital part of your engine. It is the seal between the cylinder head and the engine block, so the top bit and the larger lower part. So it is vital to be able to spot any issues before they become seriously expensive and to avoid buying a Banger with this fault. Always watch the coolant temperature. The gasket prevents the coolant and engine oil from mixing together. If they do then, the temperature will rise especially because of the hot oil vapour moving to where the coolant is meant to be reducing the temperature. If the Banger seems to overheat and the level moves into the red section or there are warning lights or messages, then you must stop the Banger when it is safe to do so. Only check the coolant when the engine is cold. So if the Banger is constantly losing coolant, it may be because it is leaking from the cooling system and mixing with oil. Check if there is oil on the dipstick, any froth on the metal may be an indication that coolant has mixed in with the oil. Now take off the radiator cap, or usually the cap on the expansion tank, which is attached to one side of the engine bay, and look at the water. White deposits point to oil and water mixing. Put your fingers in, then rub them together; do they feel oily? Look at the exhaust. When the engine is running there may well be water dripping from the exhaust as it has worked its way through the system. A garage pressure test is the definitive decider on the issue, if there is any doubt. Fixing a head gasket usually means a top end engine rebuild, provided that the engine has not seriously overheated and needs complete replacement. It is possible to buy a relatively cheap additive that seals any leaks. This may work as a short-term fix as you decide whether to sell or repair the Banger.

When does the Banger need servicing or at least, professional attention? Some would say that it is common sense. Sometimes it could just be a calendar, which tells you the Banger, is due for a garage visit, or a service light that blinks on dashboard. There are however some obvious signs, which should prompt you into making that call. When it becomes uncomfortable. It may be hard to spot this, get used to how a Banger behaves and get used to any faults. The general rule is, when the Banger starts crashing over bad roads, or bouncing it suggests that there are suspension problems. Refer back to all those noisy things we identified. When it pulls to one side under braking. This issue can also be related to problems with the suspension, but mostly unpredictable behaviour under braking suggests that the brakes need adjustment, or new pads and discs. When it starts to produce smoke/steam. Hopefully you won't miss this when it happens, otherwise you may need to visit the optician. Smoke is an engine thing and is a result of burning oil. That can mean there are many engine issues, usually related to wear and tear. Steam is usually an overheating problem as the coolant evaporates. When the engine light comes on. Never ignore, the lights on your dashboard. At the lowest level there might be a gentle reminder to take your Banger for a service via a glowing spanner. If however, there is a yellow engine check light, this is much more serious and it is essential to stop driving. There are also lots of other warning lights and possibly messages on the dashboard, the only way to put these out is to visit a garage.

Warning Lights. They usually mean that something bad and potentially expensive is happening to your motor. **Oil.** The one that looks like Aladdin's Lamp, suggests an engine oil leak. The solution depends on where the leak is; when new gaskets or seals are required then this is a garage job. However, Increased oil consumption and noticeable blue smoke from the exhaust, could be because a worn engine is burning oil. A garage may need to fit new valve guide seals, or more seriously new pistons and rings, maybe a new engine. **Airbag**. The one that looks like a beach ball and suggests that the airbag may not go off in the event of an accident. It is an instant MoT failure as well. This could be a simple sensor failure but it is essential that you contact a garage. **Engine**. If the engine or ECU (electronic control unit warning light) comes on it is likely that the engine is suffering from a lack of power or making strange sounds. If not it could just be a minor sensor failure. Either way you will have to stop driving and contact an expert. **Brakes.** Another potential serious failure. It may warn of seriously

worn brake pads or some other fault with the system. There could be a leak in the brake fluid system itself. Power Steering. There could be a leak in the hydraulic system, or a sensor fault with an electronic power steering system. **DPF Filter.** Here is an increasingly common warning light related to the Diesel Particulate Filter, which could be blocked by soot and is often accompanied by smoke from the exhaust. This could be an expensive problem so go to a garage. **Coolant.** There could be leaks from hoses, possibly the radiator. This could lead to overheating. Best to stop driving, check the levels and speak to a garage. **Battery.** Suggests that the battery levels are dropping rapidly. In most cases the battery is due for replacement or at least charging. Batteries have a lot to do these days, but usually the warning sign is just that. Not serious, but you do need to contact a specialist to get the battery checked. More on batteries in a few pages. **Tyre Pressure.** Usually this means that the tyre pressure is dropping, caused by a leak or slow puncture. It is best to stop driving until you have identified which tyre is the problem. Overall keep an eye on those pesky warning lights and never ignore them.

Banger Checks. If you want to stop bad things happening and those irritating warning lights coming on, it pays to pay attention to some of the more vital parts of your Banger. Check Engine Oil. According to the experts you should check your oil level at least every 500 miles or once a week. It is the lifeblood of your Banger. Oil lubricates your engine and stops it shuddering to an expensive halt. That's why if you care about your Banger you must check your oil level regularly. At the risk of sounding patronising you should place the Banger on as level a place as you can find. Drives and roads can have inclines and many drivers just don't allow for this. I'll assume you know what the dipstick looks like. On older Bangers it will be sited in the middle of the engine. Modern Bangers have a bright yellow cap. Take it out, wipe clean and feed it in. Then take it out. On most vehicles there are two marks on the gauge. The oil level must be between the upper and lower marking. The owner's manual reveals how much oil you must add for topping up. Generally this will be between 0.5 and 1 litre. Don't be in a hurry, just add a little at time and return after five to ten minutes to check the level. You must consult the manual or the manufacturer for the definitive procedure because although the engine should be switched off for the majority of Bangers there are some exceptions. When you add oil it is a good idea to have a clean rag ready so that you can wipe around the cap so that any dirt doesn't get inside the engine then keep it nearby to mop up any spills. Don't add too much oil it sounds obvious, but you should never overfill an engine. That's because seals may start to leak, foaming of the oil occurs which can push oil into areas where it is not supposed to be. And finally always use the correct oil and grade. The owner's manual or the Banger's manufacturer will be able to tell what that will be. Any increase in oil consumption could be engine oil leaks from old seals and gaskets. Blue smoke from the exhaust means a worn engine is burning oil.

Bodywork. Bangers don't rust anymore do they? Well actually they do. Not as badly as older motors used to, it can though still be an issue and if serious can result in an MoT failure. The trouble is rust is very good at hiding away, so it is vital to go and find it and get rid of it. The nasty little rust bug, starts at the bottom of the Banger where moisture and water collects and then eats its way upwards through the bodywork. What you have to distinguish between is serious and superficial rust. Look underneath the Banger without jacking it up and use a torch. Inside, pull up the carpets especially if they seem to be damp. The area adjacent to the doors is the most common area for rust. On the bodywork it should be really obvious because of bubbling and discoloured paintwork. Some Bangers may have plastic or aluminum panels which don't rust. The Banger may be full of holes but it can still be perfectly legal. Generally, minor rust on the edges of panels is rarely serious, so a rusty bonnet, doors and wings can easily be lived with, or simply repaired. Structural rust is more serious, can make the Banger unsafe and will result in an MoT failure as the important load bearing areas of the Banger can no longer take the strain. The problem is that whereas most mechanical parts can be easily and cheaply replaced; properly fitting and then painting body panels is expensive.

Tyres. Yes I know we've covered tyres when buying, but they are really important, especially when you are responsible for them on a daily basis. When was the last time you paid any attention to those four rubber rings at each corner of your Banger? Probably when the MoT man shook his head and pointed to the lack of legal tread, or when you pulled onto the hard shoulder and looked forward to changing a flat tyre in the rain. Indeed, that's exactly why we should all be fascinated by tyres because they are probably the four most important components of any Banger. The essential ingredient for keeping tyres in good condition, to provide maximum performance and give maximum life is air. It is air that keeps tyres fit and safe and carries the weight of the vehicle and its load, not the rubber or the casing material as many people think. Tyres should be checked regularly, at least weekly either at home or when visiting a garage. Examine them for obvious signs of under-inflation, wear, cuts in the tread or sidewalls, bulges in the sidewalls and stones trapped in the tread grooves, which should be removed. It is also useful to check for leakage at the valves, especially following inflation, and to replace missing valve caps. If tyres show any sign of uneven wear the vehicle should be checked to ensure correct alignment and balance of the wheels.

Tyres on a Banger should also have a minimum of 1.6mm of tread across the central three-quarters of the breadth and in a continuous band around the entire circumference of the tyre. Tyres must also be suitable for the use to which the vehicle is being put such as towing and inflated to the correct pressure. Tyres of different types must not be fitted to opposite wheels on the same axle.

Tyre pressures should not be checked during or immediately after a journey while the tyres are still warm. This will result in an incorrect pressure reading. Ideally buy a pencil-type pressure gauge so that tyres can be checked regularly and kept in your pocket, or elsewhere. Garage forecourt ones can be inaccurate. Follow the manufacturer recommendations at all times and when in any doubt take your Banger to a tyre expert at a garage.

Battery. According to experts the number one reason why your Banger won't start is that the battery is flat. So it makes sense to keep an eye on this most vital and easily maintained component of your Banger. Batteries go flat because of lack of use, or a faulty cell. Most batteries are unlikely to last longer than five to seven years although there are exceptions. Apart from age and lack of use, cold weather affects the performance of a battery, especially as it is required to power so many functions from in Banger Entertainment (radio) and wipers to lights. Sluggish starting, headlights dimming suggests that the battery may be nearing the end of its life and not holding its charge. Ideally, never attempt to start your Banger with lots of gadgets switched on. Also any erratic electrical behaviour points to a possible problem with the battery. Although you can go to a garage to get a battery checked, you can use a test meter, or a battery charger with a display to see what the condition is. Usually the reading should be above 12 volts, below that, the battery may be flat. Look after your battery by checking the terminal connections (the leads bolted to the battery) for tightness and clean away any corrosion either with a brush, or just warm water. A layer of Vaseline around the terminal posts will keep the white corrosion deposits away. Do not forget to check the security of the battery that may be held in position with a strap, or a metal plate. If the battery falls over, it could start a fire and cause considerable damage to the surrounding area from spilled acid. If you only do short journeys with long gaps in between this is not good for a battery so you should buy a trickle charger to maintain its condition.

Lights. Making sure that your headlamps, indicators and fog lamps work is one of the basics. If they don't it makes you less safe, open to prosecution for having an unroadworthy vehicle and an instant MoT failure. Being able to see and be seen is a Banger driving fundamental. Seems obvious, but the simplest thing you can do is to keep your lights clean. This is especially important in winter with the build up of salt and road dirt. Ideally clean them once a day or every few days. Also look at the plastic covers, which many vehicles have over the headlamps. These can, over time, discolour and will eventually become an MoT fail point. You can buy a lens cleaner, but what also works is something like toothpaste which is mildly abrasive. Check your lights each week. Get a friend or family member to double check brake lights and indicators, otherwise shop windows or other reflective surfaces may help. It isn't always easy to tell whether a bulb has blown. You may have a vehicle with warnings that show up on the dashboard. Otherwise it could just be other drivers flashing and waving their fists at you. Ideally you should be alert enough notice reduced visibility. Often indicator bulbs tick much more rapidly when one of them has failed.

There is no standard procedure for replacing lights on any vehicle as it now varies so much. You can of course get a garage to do it or sometimes a car parts supplier will fit them, for a price of course. So it is worth reading your owner's manual and even more usefully there are a lot of videos online showing exactly how it can be done. Some experts think that bulbs should be replaced in pairs, for more consistent performance. Also, some models may even need a computer reset to acknowledge that the bulbs have been replaced. You could upgrade to better bulbs and research is essential to make sure you find compatible, but more powerful bulbs. The budget option is halogen bulbs, which claim to emit at least 30% and as much as 100% more light. More expensive and with a longer life, up to ten years, are the HID (High Intensity Discharge) or Xenon bulbs which can put up to 120% more light on the road compared to a standard halogen bulb. Xenon can also produce a beam pattern that is longer allowing other road users to be seen more clearly and giving you more time to react to potential hazards.

And finally, one of the most important and unanswerable questions is, **How long should a Banger last?** Much depends on how the vehicle is used and driven. Bad driving will always cause parts to wear out faster. Also different makes and models deteriorate at varying rates. This is Bangernomics, so let's have a go at answering the impossible by looking at various bits of your Banger.

Tyres. Manufacturers can't put a finite life on tyres as there are so many variables. Usually a branded tyre from a well-known manufacturer will last longer compared to a cheaper option. The most important thing is knowing when to replace ones that could become illegal. Lifespan is not as important as performance, when it comes to comfort and how they make the Banger handle. Hard driving cuts tyre life right back especially if the driver spins the wheels and slides around corners. In that case they will struggle to get 10,000 miles from a tyre. Most of us should expect around 25-30,000 miles, but it could be more, or much less. Tyres, just like us, can simply get old and tired. In certain weather conditions they could begin to crack after a few years use. So even if the tread is perfect, the tyre could be illegal and unsafe.

Brakes. Again, this is very dependent on how a Banger is driven. Brake pads could last 40,000 miles, with the fronts wearing much faster than the rears, which do not have the weight of an engine on top of them. However, stop start driving will increase wear. Also the type of vehicle is important, as a large 4×4 will put a lot more stress on the brake parts. The brake discs should also last at least 50,000 miles. However it does depend on the quality of the parts fitted, cheap makes could need replacing every year.

Drive Belts. There is often a manufacturer replacement schedule suggesting when a fan or other drive belt is replaced. This can be every two to four, years, or when a vehicle reaches a certain mileage. Usually these belts will last 30-40,000 miles. The most important is the timing belt in the engine, which cannot be visually inspected. Whatever the manufacturer states should be the replacement period should be followed. Usually it is at least 5 years or 50,000 miles.

Shock Absorbers. Yet again this depends on how the vehicle is driven and also the conditions it has been driven in. They would be expected to last at least five years and 50,000 miles. They should always be inspected for leaks or any other damage at every service.

Rubber parts. Often it depends where they are. If not exposed to sunlight then the hoses under the bonnet on the radiator, fuel lines and other connections could last a decade. They will eventually dry out and crack.

Plastic parts. These days many engine parts are made out of plastic such as intake manifolds and covers as well as the interior trim, dashboard and seats. In theory, they ought to last the life of the Banger. However, depending on the conditions, they can become brittle and crack.

Exhaust. It depends on the system fitted and they have to put up with a lot of with water and corrosive acids running through them. Often it is sections of the exhaust that fail first after about five to seven years. Steel systems can struggle to five years, whilst stainless steel usually last for over a decade. It depends on how and where the vehicle is used. Salt used on the roads during winter can shorten the life of a steel exhaust significantly.

Gearbox. Manual transmissions will often last the life of the Banger. However, the clutch depends completely on how the vehicle is driven. It can be burnt out very quickly by a sporty driver, or someone who rides the clutch in traffic. It is possible to make a clutch last 60,000 to 100,000 miles or longer. Automatic gearboxes are harder to abuse, but more complicated to repair. They should still last well over 100,000 miles.

Pumps. There is usually a fuel and a water pump in a Banger and sometimes they can last the life of the vehicle. After 100,000 they could potentially fail and could also have a much shorter life. Fuel pumps are usually located inside fuel tanks and can be expensive to replace.

Battery. A lot depends on the conditions under which the Banger is used. Modern Bangers in particular put huge demands on the battery and after 5 to 6 years its performance will start to tail off. Also extreme heat and cold put extra stresses on a battery, so potentially it could last less time. Check if there is a guarantee you can claim on.

Engine. Often can last the life of Banger if well maintained and serviced properly. With regular oil changes it has been possible depending on the type and make of vehicle to cover 300,000 to 500,000. It will be the gradual replacement of parts, which will help maintain their optimum condition. Staying at proper operating temperature, long runs, rather than lots of short journeys help keep an engine healthy. Also a petrol unit has less stress and should last longer.

Anti Theft...Before I forget, ideally you want to keep hold of your Banger. There are plenty of simple things you can do and if by some fluke you have a sophisticated Banger or are reading this in 2045, then you'll need a special cover for your remote keys. A simple RFID wallet can prevent your fob being hacked by a thief intercepting the electronic communication between it and your Banger. The trouble is, once inside a thief who is determined to take your Banger and either has stolen your keys or may be able to access the electronic control port inside your car. That's why you always need an old-school steering wheel lock. These are basic, but will slow down most thieves or just make them think twice and move onto another vehicle. Even more of a deterrent would be a wheel clamp. These are a bit more expensive and always make sense if your Banger is left for long periods in the same spot. Hardcore Bangernomicists, (see a few chapters on) will fit what are called, kill switches. That is a hidden device which stop the engine starting. Just in case you wondered it will just turn over by interrupting the power supply to the ignition. Otherwise, locking your Banger is a great thing to do, plus there are many devices you can buy. Perhaps the best advice is to park your Banger in a garage, if you have one. Some people have them and they are full of rubbish, so what a great excuse to get rid of all that nonsense. Ensure it also has an alarm system and if you are flush with cash, or possibly have a Luxo Banger, a close circuit television system might help. At the very least make sure the area around your car or garage is well lit.

Look after your Banger and it will look after you. This is pretty much all you need to do. It isn't complicated, just keep on top of any issues and you will find that long term it will be cheaper to run and easier to look after.

Ask friends/relatives for recommendations. Are they members of a professional body? Ask manufacturer for repair times and what is actually serviced. Tell garage in writing what you want done Check whether costs for labour, parts and taxes are all included. Ask if workmanship and parts are guaranteed. Get quotes and second opinions from other garage Request that you be consulted about any additional work/expenditure. Agree collection and payment arrangements. Get a detailed invoice showing a breakdown of all parts/labour costs.

GARAGE CHAINS Rather than being a family owned garage there are a number of nationwide chains, sometimes called fast fit centres with lots of bays and ramps and the promise of fixing your car for less. This is not always the case. It is vital to compare prices as often a local garage can match or beat a quote. However, these chains may be best for supplying and fitting some consumables, like tyres and exhausts.

WEEKLY CHECKS: What you have to do is get into some sort of routine. Put aside fifteen minutes on the weekend just to carry out these simple checks. It means you will be able to quickly spot any deviations from the norm, so that you can trace any fault and mostly importantly get it fixed. The following are general recommendations and are no substitute for consulting an expert mechanic, which you must always do when in any doubt. Also many cars have diagnostic systems and could very smugly tell you what is wrong. However, the sensors located at various parts of the car can become corroded and dirty, so are not completely reliable.

OIL: According to the experts you should check your oil level at least every 500 miles or once a week. And at the risk of sounding patronising you should place the car on as level a place as you can find. Drives and roads can have inclines and many drivers just don't allow for this. I'll assume you know what the dipstick looks like. On older cars it will be sited in the middle of the engine. Modern cars have a bright yellow cap. Take it out, wipe it off and feed it in again. Take it out and now read it. This is necessary; as it's likely some oil will creep along the gauge when the engine is shut off so the first measurement will be faulty. On most vehicles there are two marks on the gauge. The oil level must be between the upper and lower marking. You will find in your owner's manual how much oil you must add for topping up. Generally this will be between 0.5 and 1 litre. Don't be in a hurry, just add a little at time and return after five to ten minutes to check the level. You must consult the manual or the manufacturer for the definitive procedure because although the engine should be switched off for the majority of cars there are some exceptions. When you add oil it is a good idea to have clean rag ready so that you can wipe around the cap so that any dirt doesn't get inside the engine then keep it nearby to mop up and spills. It sounds obvious, but you should never overfill an engine. That's because seals may start to leak, foaming of the oil occurs which can push oil into areas where it is not supposed to be. And finally always use the correct oil and grade. The owner's manual or the car's manufacturer will be able to tell what that will be.

PROBLEM: Noticeable increase in oil consumption when checking level and patches of oil when parked. **CAUSE:** Engine oil leaks. **SOLUTION**: Depends on where leak is; when new gaskets or seals are required then this is a garage job.

PROBLEM: Increased oil consumption and noticeable blue smoke from the exhaust. **CAUSE**: Worn engine is burning oil. **SOLUTION**: Garage may need to fit new valve guide seals, or more seriously new pistons and rings, maybe a new engine. Oh no!

HYDRAULIC FLUID: Both the clutch and brakes rely on specialised fluid to keep them working. Sometimes they are located in plastic bottles and at other times metal pots towards the back of the engine compartment. Refer to your manual for the correct levels. Only a tiny amount of fluid is necessary to maintain the correct level. Do not allow the level to drop as air could enter the system and seriously affect their efficiency. Be careful when handling the fluid, as it is highly corrosive both to paintwork and skin. If you need to top up then unscrew the cap, remove it carefully and place it on a piece of clean cloth. Top up to the max mark and use a good quality fluid.

PROBLEM: Brake fluid level drops rapidly. **CAUSE**: Leak in system. **SOLUTION**: Don't drive the car and consult a garage.

POWER STEERING FLUID: Check where the fluid reservoir is in your owner's manual or service book. If you need to remove the filler cap, wipe around it and then remove. Some may have hot or cold level markings and depending on the temperature ideally the fluid should be at that level. If it needs to be topped up, make sure you use the correct brand and type of fluid. Don't overfill and screw the filler cap on tightly. **PROBLEM**: Fluid level drops. **CAUSE**: Possible leak in system. **SOLUTION**: Get it checked at a garage.

BELTS: One or more of these usually runs around several pulleys (such as crankshaft, alternator and fan) and probably will be connected to other engine parts such as air conditioning, power steering and water pumps. They should be at the correct tension so that they either don't slip, or aren't on so tight that they strain the components. Push and pull the belt where it has the longest connection, the movement should be no more than 5 to 10mm. For adjustment, consult the manual, but the usual method is to slacken the bolts on the alternator and move it until the tension is correct. Now look at the belt for signs of fraying or cracks, any shiny areas are bad, as are splits. If you have any doubts about the condition of belt (which normally lasts about two years) then replace it immediately. **PROBLEM :** Screech or squeal, the belt may be too slack. **SOLUTION:** Tighten. **PROBLEM**: Constant hum at idle or when pulling away. **PROBLEM**: The belt may be too tight. **SOLUTION**: Slacken.

COOLANT: Some cars have sealed cooling systems, which do not require regular attention, but it is always worth checking the water level, as you don't want to overheat. Never check when the engine is hot and, if warm, use a cloth to remove the radiator cap. Refer to manual, but usually the water must be visible and covering a level mark. Plastic expansion tanks are also common and will usually be marked high and low on the side. This is also a good time to check for anti-freeze especially if it is getting towards winter. If there are no greenish traces, buy some from an accessory shop and follow the instructions. Consult the manufacturers handbook on this one because cars now have sealed for life systems or must use certain brands or types of fluid.
PROBLEM: Water level drops rapidly. **CAUSE**: Leaks from hoses, possibly radiator. **SOLUTION**: Replace hoses, fit, or repair radiator. Radiator additives to stop leaks are available from accessory shops. **PROBLEM**: Water loss, but also noticeable overheating when the temperature gauge indicates high or is in the red. **CAUSE/SOLUTION**: If there are squeaks and leaks from the water pump then that may need replacing. The radiator could be blocked, or leaking, then stop leak with additive, clean, or replace.

BATTERY: Most batteries are now sealed and there is no need to check the water levels. There may be condition indicator and you may need to consult the handbook or instructions to read it. However some older and classic car batteries require you to check the levels, keeping the water just above the tops of the cells, use tap water, or distilled water which can bought from accessory shops. Check the terminal connections for tightness and clean away any corrosion either with a brush, or just warm water. A layer of Vaseline around the terminal posts will keep the white corrosion deposits away. Do not forget to check the security

of the battery, which may be secured in position with a strap. If it falls over, it could start a fire and cause considerable damage to the surrounding area from spilled acid.
PROBLEM: Battery levels drop rapidly. **CAUSE**: Charging rate high, or the battery plates are faulty, or a cell could be leaking due to a crack in the casing. **SOLUTION**: In most cases the battery is due for replacement. Talk to a specialist if in doubt, though some leaks can be repaired with a special filler. **PROBLEM**: Sluggish starting. If headlights dim or go out when starting. **PROBLEM:** Battery may be nearing the end of its life and not holding its charge. **SOLUTION**: Use a test meter, or a battery charger with a display to see what the condition is. Usually the reading should be above 12 volts, at 12 or below the battery may be flat. If you only do short journeys this is not good for a battery which may need more regular charges.

HOSES: Check the ones attached to the radiator, the top hose is the easiest to see. Are there any obvious water leaks, especially from the ends? If so are the clips holding it in place tight enough? Use a screwdriver. Are there any cracks or splits? Now squeeze it, if lots of cracks appear or the rubber feels very soft then it must be replaced. Usually this is a simple job and will save you much grief. Look also at the rubber pipes, which connect to the heater, or air conditioning system, which should also be in good condition. **PROBLEM:** Water level drops rapidly. **CAUSE**: Leaks from hoses. **SOLUTION**: Replace hoses, fit, or repair radiator. **PROBLEM**: Water loss, but also noticeable overheating when temperature gauge high. **CAUSE/SOLUTION**: If the ignition light stays on then the fan belt needs replacing.

SCREENWASH: Usually the plastic screen wash container is fairly obvious either to one side or rear of the engine bay. You can simply top up with tap water although sometimes that can be a little be harsh and contain lime scale. If you really care use filtered or softer rainwater. Whatever you do though put a proper additive in, (not washing up fluid) which will de-ice and help clean the screen. Some cars have one container others may have separate ones for headlamp and even rear washers. When filling up wipe away dirt from around the filler neck, then pull off cap. **PROBLEM**: Falling level even when screen wash not being used.

CAUSE/SOLUTION: Possible split in container or in pipes. Replace broken item. **PROBLEM**: Blocked washer jet. **SOLUTION**: unblock with a pin.
WIPERS: Look at the blades, they must not be perished or frayed. Ideally give them a clean as detailed before, but you can also clean the edge of the blade with a cloth dipped in screen wash. **PROBLEM**: Areas were left unwiped, oily streaks which restrict view. **CAUSE**: Windscreen wipers are perishing and may not be making proper contact with the window due the mechanism being worn. Dirty windscreen. **SOLUTION:** Replace windscreen wipers and add special screen wash to the washer bottle.

TYRES: Check their pressures ideally with a pencil type gauge, which you push into the tyre valve. Take readings when the tyres are cool, otherwise the pressure increases and gives a false reading. If you overfill release air by pressing loosely against the valve. Use either a foot pump, or forecourt inflator (which usually has to be paid for). We should also look at the car's handbook (also in inside of fuel flap) for the definitive information on tyre pressures front and rear and variations when there are just two passengers as opposed to five. Also check the condition of the tyres for wear, cuts in the tread or sidewalls, bulges in the sidewalls and stones or foreign objects trapped in the tread grooves, which should be removed. It is also useful to check for leakage at the valves, especially following inflation, and to replace missing valve caps. Don't forget to check the spare (if fitted) too.

PROBLEM: Tyre pressure dropping. **CAUSE**: Leak or slow puncture. **SOLUTION**: Consult tyre specialist as it may be beyond repair. Always get a second opinion if the fault is not obvious enough to you. **PROBLEM**: Uneven tyre wear, especially on the front wheels. **CAUSE**: Misaligned brakes, or damaged suspension. **SOLUTION**: May simply need the tracking corrected, which is a cheap, and simple garage job. Could be suspension, which is more dangerous, consult a garage. **PROBLEM**: Worn centre section of tyre. **CAUSE**: Over inflation. **SOLUTION**: Correct pressure.

LIGHTS: Don't wait for a car's warning light system to tell you that something may not be working. They aren't always

accurate anyway. An incredible amount of muck is thrown up from the road, which then clings to the light lenses. This means other drivers won't see you and in turn you may not be able to see where you are going. So simply wipe them over with a damp cloth or sponge. Periodically check that all the lights work with a friend doing the honours on the outside. Alternatively, if you haven't got any friends then the reflections from shop windows can provide an instant check.
PROBLEM: Indicators flash faster than normal on one side of the car. **CAUSE**: A bulb has blown or one that is the incorrect wattage has been fitted. Consult your manual.
PROBLEM: Indicators flash very slowly. **CAUSE** incorrect bulb, faulty electrical circuit, faulty flasher unit. **SOLUTION**: Replace bulb or indicator unit, or clean up terminals see below. **PROBLEM**: No indicators. **CAUSE**: Blown Fuse. Wiring Fault. Indicator switch broken. Flasher unit broken. **SOLUTION**: Replace units, but essentially you may need the whole circuit to be checked. **PROBLEM**: Light does not work, or only works intermittently. **CAUSE**: Bulb may have blown look at the wire element inside the bulb. Otherwise look at the connections for any corrosion. **SOLUTION**: Clean connections and use a water dispersant (WD40). Look at wires for security and check the earth wire, which connects to the bodywork, is tight and not corroded. Again clean, use sand paper and water dispersant.

ELECTRICAL FAILURE: If something doesn't work (radio, windscreen wipers) then more than likely a fuse will have blown. Replacing them is easy. Refer to the manual to locate the fuse box and simply pull out and replace with one that has an identical rating. If it blows again then there is a problem with the electrical circuit or the item itself.

CLEANING: Something you ought to do at least every few weeks or months. It means that you get to examine the bodywork closely and can spot any problems or faults, or leaks more quickly.

DURING THE WEEK: Simply make a mental note during your weekly motoring trips of any strange noises, or driving characteristics. If the brakes are less effective or the engine is sluggish then you will need to get that inspected. Refer back to the BANG, SPLUTTER, CLANG chapter, or consult your local garage to identify the problems.

REPLACEMENT PARTS

You really do need to think Bangernomically and that means not taking the simple option when it comes to parts. You could just visit a nearby main dealer, who may well be the cheapest and possibly the only supplier. However that is doubtful and many manufacturers end up deleting certain items from their stocks eventually. So you will have to resort to alternative sources and with the global marketplace you could end up buying from a private seller, a specialist parts supplier, an owner's club or a dealer on the other side of the world.

Body Panels: These can be bought second hand and cleaned from a salvage or scrap yard. There are also 'Pattern Parts' effectively low priced copies which are actually more difficult to fit sometimes, but perfectly adequate.

Engine/Gearbox: The oily parts of a car are usually terrifically expensive when bought new. However, there are companies which recondition these parts. You give them your worn out bits and they exchange it for one they prepared earlier.

Interior: Pieces of trim and even seats can be expensive and impossible to find, so it is always best to go to a salvage yard first. Owner's Clubs are great places for these sorts of obscure items.

Banging About – What can you do?

Just Keep on Banging on and on and on

Simply keeping up with the maintenance schedules, is all you have to do. If you have never done this before or you have not bought your Banger yet then practice on someone else's car. Ask if you can look after it for week. You never know you might save a fortune by tipping them off about an impending problem. Now that really will impress them.

12. Bye Bye Banger:
Banger disposal

Selling your Banger is all part of the process and for many is the least enjoyable and most hassle. Keep in contact with other cheap car owners and they might be able to help when it comes to the crunch. The best thing about disposing of a Banger these days it that it won't cost you a bean. It just takes a bit of creativity and determination and you will turn that Banger into Beans (Money).

When should you repair or replace your Banger? This is the most difficult Banger owning conundrum anyone has to deal with. Usually it boils down to this: should you fix or flog? Sell your Banger when the repair costs are more than the Banger's value. Here is the easy decision. Simply check the value of your Banger in useable condition, compared to what you have been quoted for repair. If the repair itself is far more than the Banger is worth, then you are better off selling. What has to be factored in is the additional hassle of finding and buying another equivalent Banger. Ask the garage who quoted for the repair, whether there are any other issues. If not and you like the Banger, then keep it. Bangernomics is all about keeping a Banger running and the last thing you should worry about is how much it will fetch in the future.

Condition: The best time to sell is before the car needs major repairs, so pass the crystal ball. By now you might be able to recognise the tell tale sounds of expensive repairs. The principle is the same when buying, serious body repairs are costly, mechanical items can usually be simply replaced. A replacement reconditioned engine may well be cheaper than the hassle of selling the car you have and buying a replacement, which will be an unknown quantity.

Time: Buying and selling cars is largely a seasonal activity, Obviously the right car at the right price will always find a buyer, but generally, don't sell in mid-winter, or after the spring and autumn new car rush when there are lots used vehicles about. In high summer everyone is on holiday and you may not get a good response to your advertisement whereas spring is a traditionally popular time for people to go shopping for cars. Often if a car is about to expire you don't get much choice about the timing of the sale you just need to get rid of it.

Can you afford it? Doing your sums again, what is the shortfall between what you can expect for your car and the amount you want to spend on the replacement?

When to sell your car is one of life's great imponderables. The exception is when a mechanic shakes his head and the garage quotes an eye-watering sum for helping it to limp through another year. Alternatively you might need a different kind of car one that's bigger, smaller, faster, or more frugal. As you must have gathered by now Bangernomics shows you how to buy and run a car cost effectively, so to make this a comprehensive guide to selling as well, would mean that the book would be twice as thick and twice the price. However, turning your car into cash is a very important principle of Bangernomics. So when should you consider selling your Banger?

Sell your Banger when the mileage is above average. The truth is that a well-maintained Banger can last for many 100,000 miles, so the number of miles it has covered is irrelevant. Many used car buyers get frightened by high mileages. In theory the average annual mileage is around 12K a year. So a five-year-old Banger should have 60K miles. Just to recap, mileage is always irrelevant if it is working. High miles means less demand, that's all. Don't worry, just be grateful that your Banger is reliable and running. Sell you Banger when it is rusty One of the most expensive parts to repair on any Banger is the bodywork. Whereas mechanical items can often be simply unbolted and replaced, anything to do with the body structure involves more skill and complication. Painting and finishing to a decent standard takes time and money.

Sell your Banger if there are tax or insurance changes. Perhaps a certain size or type of engine is due to be re-taxed under proposed government changes. Perhaps your insurance company raises your premium and you cannot get a more competitive quote elsewhere, then it could be time to change the type of vehicle you own. Your Banger might suit someone else's lifestyle or budget better.

Sell your Banger if you want a change. You knew that one already, when you're a bit bored... So you actually want to sell your Banger? These days it could not be easier to put your Banger up for sale online. In seconds your Banger is out there for the world to see and buy. However, if you haven't prepared your Banger properly, buyers won't just be disappointed, they will see lots of reasons not to pay your asking price. So what are the essential things you should do?

CLEAN IT! It's a fact: scruffy and smelly cars don't sell. Now cleaning the car you are about to sell sounds really obvious, but many forget to do the basics. Make the time to give the car a clean and not only will it will be easier to sell you may be able to ask for and get a higher price. To buyers a car that is clean looks like it has been cared for.

So without going into huge detail simply vacuum and polish the interior then wash and wax the bodywork. Of course there is the option of getting the car professionally valeted. That saves you time and elbow grease, but it eats into your profit. If you are lazy just go down to the local car wash and the jet wash is money well spent. Fine detailing on a classic or luxury car may be helpful, but on the vast majority models it really isn't necessary. Also a car that is over prepared can put buyers off. Indeed, too much polishing will actually make scratches stand out. However, cleaning the car yourself makes you aware of exactly what needs to be fixed.

BODYWORK Minor car park scrapes and thumbnail sized dents are not a problem, especially if the car is over five years old. As it is a Banger you can get away without spending on what will be expensive bodywork repairs. Big turn offs for buyers (well those who don't understand the whole concept of Bangernomics) are missing bits of trim and door mirrors. A trip to the salvage yard to pick up these items is absolutely essential. Make sure that all the doors and tailgate open and close easily. A squirt of silicone lubricant will shut up any squeaks and the odd buzz or flap could be some loose trim, so tighten the screws. **DO** Fix and or replace broken trim. Make sure doors can be opened easily/quietly. **DON'T** Buff black bumpers to a high gloss, or over polish bodywork or worry about parking/supermarket dents, it's a used car.

ENGINE Every buyer wants to look under the bonnet even if they don't know what's going on in there. Buyers expect to see a tidy but honestly grimy unit. Don't be tempted to pressure wash because you could kill the electrics and you put buyers off who are suspicious of a shiny engine, it may even suggest that a serious oil leak is being hidden. By all means clean the oil filler cap and water expansion tank. Those are the things a buyer will touch and check and it looks like the previous owner did that too. Indeed why not change the oil? Make sure that the battery is secure and not caked in corrosion. Oh yes the bonnet stay should be there, or the hydraulic supports should work as it would be a bad idea for a buyer to bump their head. As for the engine, well if it isn't broken, don't even attempt to fix it. Ideally it should start first time, run evenly and quietly whilst squeaks can often be traced to loose belts. **DO** Check fluid levels, battery condition, bonnet opening, make sure engine running smoothly.

DON'T Pressure wash, or brighten up by polishing under the bonnet.

INTERIOR Like we said, smells don't sell. If the car has been home to a smoker or a dog then you will have to work hard to get a neutral aroma. Vacuum and scrub away at the ashtray, clean head lining and wash or even replace the floor mats. Clean the dashboard, but don't be tempted to get a gloss finish. Overall, depersonalise the car because people don't want to see holiday stickers and football team badges, or what you enjoy snacking on. The boot may have more stains so use a carpet cleaning solution and take the opportunity to make sure that the tool kit is there and that the spare is usable and inflated. **DO** Get rid of smells, personal items and use domestic carpet cleaners. **DON'T** Buff dashboard to a high shine or leave you personal effects lying around.

WHEELS/TYRES You can be certain that buyers will kick these, so look at their condition. Alloy wheel refurbishment is costly, but lots of parking scrapes and chips on a very young car looks bad. Otherwise they should just be clean. Missing wheel trims instantly make any car look scruffy and should be replaced, often you can find them by the side of the road, but they are unlikely to match. Even the least clued up buyer can spot illegally worn tread and you must not sell an unroadworthy vehicle. Worst of all don't cheer old tyres up with a lick of tyre paint, it looks very tacky.

DO Replace illegal tyres and replace missing wheel trims. **DON'T** Use tyre paint.

GET ORGANISED You have got the car sorted, but one of the most crucial parts of used car preparation is getting all the documentation in order. Indeed you can't register a car without the registration document now and any used car without this piece of paper can become worthless. Buyers like to see the paperwork neatly presented and not in a chaotic torn and tatty envelope. Get a clean file and clip in the registration document, the most recent roadworthiness certificate if applicable, the owner's manual in a clear wallet, plus any invoices for work done and especially any relevant warranties.

Value your Banger. Finding this out can be something of a challenge. However, it is vital if you want to get the best price for your vehicle when it comes to selling it privately or even part exchange against a replacement Banger. The easiest way to check what the trade price of your Banger is to enter the vehicle details into an **Online Buying Service.** Interestingly, this is what car dealers often use to give a quick bid on your Banger. They know this is the place that most sellers will go eventually to check the price. Make sure you enter all the details accurately and get used to this being the going rate for your Banger.

Auction Website Another way to test the marketplace and get the going rate is to put your Banger on an auction website. Here is probably the easiest way to find out what someone will actually pay for your old Banger. Listing it with a 'reserve' amount, so you don't lose out, and even with a 'buy it now' if you don't have the confidence. However, many find that a no reserve auction with 99p start can convert into increased interest and a high final bid. Remember, this is an auction so anything can happen.

Writing a Banger advert. These days selling a Banger could not be easier, you just stick it and some pictures on a social media site and wait for a text, a What's App or whatever the current mode of communication is. However, there is an art to writing effective motoring classified ad which will stand out from all the really dreadful ones, which will be most of them. It is crucial to get the wording right, otherwise you won't get a response and if you do, it won't necessarily be the right sort. The purpose of the advertisement is not so much to fully describe the Banger, but to get interested parties to contact you and buy.

The 10 most important elements are:

Year registered: So the buyer knows exactly which model.

Service history: If you've got it, flaunt it. 'FSH' (full service history) is an accepted abbreviation.

Mileage: Be specific if the car has a low mileage. Mileage becomes less of an issue as the car gets older and can be I dropped if you are flogging a banger.

Model: This is vitally important in a marketplace that places emphasis on differences between an L and GL. Buyers also want to know the engine size, so 1.6, 2.0, diesel and whether gearbox is automatic. Don't waste words on the spec if it costs you more, if not list absolutely everything.

Colour: A crucial buying factor to many people, a picture ad may be too small and blurry.

Condition: Honesty really is the best policy. If there is a major fault, then mention it. 'Good condition, but gearbox needs attention hence...' that lovely word 'hence' qualifies the defect and justifies a lower than normal price.

Price: The word 'offers' or 'ono' (or near offer) at the end of the description is wrong it indicates that you are desperate to sell at any price. Buyers always make an offer anyway.

Contact Details: Sounds obvious, but a phone number, email or whatever form of communication we may use in the future which may include telepathy, is always going to help.

Number of Owners: If you have been the only custodian, or there have only been a couple in the last decade, this is another reassuring indicator for the potential purchaser.

Warranties: Any existing guarantees, or warranties for parts, or recent repairs.

DEALING WITH BUYERS
Like it or not, the grumpy so and so kicking the tyres is a customer. Treat him/her as such and you stand a better chance of selling the car, but always stay in charge of the situation. Answer any questions on the phone or by email, or in person directly and don't flannel. If you are honest and straightforward you are more like to hook a potential buyer. When it comes to making an appointment don't allow yourself to be trapped into an inconvenient time, or become a prisoner in the house waiting for someone to put in an appearance. Always take the enquirer's contact details and then call back to make the final arrangements so you know they are genuine. Most of all never make the appointment late in the evening, or agree to take your car to a venue that could leave you vulnerable. Ideally if you are the nervous type or just concerned, ensure that a friend or relative will be with you at all times.

When the buyer arrives be friendly and helpful and because you are not desperate to sell this car, don't fuss around the buyer, or talk too much. Simply ensure that all the doors are unlocked, raise the bonnet and leave the buyer to it, but don't leave the keys with them! Then retire gracefully to prod at the garden, stay in the vicinity, but let the buyer look around. Don't offer to do anything, like start the car or go for a drive. If they are interested then they will ask, if not they may simply go along with your offers to keep you happy and that is a waste of everyone's time. Also, leaving them with the car makes it less likely that they will criticise it and put you on the defensive.

TEST DRIVE If the buyer wants a drive the most important thing to remember is that you should remain in control at all times. Often your insurance can cover other drivers provided you give your permission, but do you honestly want to jeopardise your policy? Make it clear that if someone wants to drive, they have to be comprehensively covered for your vehicle and you will have to see a valid insurance policy and driving licence as proof. Remember if someone doesn't have the right insurance cover, or you just don't fancy them driving, don't ever let them.

THE DRIVE Make sure that you drive first. This is so you can show that the car can be driven smoothly and there are no problems with a crunchy gearbox, juddering brakes and sluggish performance. Drive to a quiet road invite the potential buyer to take the wheel. But before they do that, take the keys with you, or they could take the car. Make sure that they are comfortable. Explain the controls. Specify a local route, avoid built up areas and school run snarl-ups. Chat freely during the drive, point things out that you think are relevant. Don't babble on. Silence often provokes favourable responses from the driver.

AFTER THE DRIVE Remain in the car and ask if the buyer has any questions and as they have looked, touched, prodded and driven it is not unreasonable to ask whether they want to buy. If they do, go somewhere more comfortable like your house and make a cuppa.

THE DEAL Be polite, low-key, helpful, never get excited and never show that you are desperate. Being chatty ought to help when it comes to the negotiations. If the buyer makes an offer be prepared to drop in price to half of the difference, but not until you have been asked, never give anything away for nothing. Don't babble because this gives away too much information and weakens your position. Always link your acceptance of any offer with a condition, such as payment within a set period. If the buyer does not bite then tell them that's the deal take it or leave it. Being polite means that the door remains open if the buyer wants to reconsider. If the buyer is keen to buy, but does not have the funds then get a deposit in cash. It concentrates the buyer's mind wonderfully stops them having second thoughts, or simply walking away just leaving a promise behind.

WEAK? If you think that you are a weakling when it comes to negotiation then why not nominate a friend, or hard-nosed spouse to do the difficult bit? After showing the buyer around the car and doing the test-drive, then 'hand over' the buyer with an 'I've got no head for figures, my friend/spouse/dog handles all that'. This is a technique used by showrooms, which softens up customers with a nice salesman and then squeezes out the cash with a nasty one.

GOLDEN RULE Never stop taking enquiries until you have been paid in full for the car. The potential buyer could drop out and you will then have a list of numbers from people who have responded to your brilliantly worded advertisement, to fall back on. Never under any circumstances accept more than one deposit, as you will get yourself into legal hot water and possibly a black eye.

SOLD! Once you have shaken hands, the hard part is getting the money. You should also put the agreement into formal terms, even though an oral contract usually exists once you have agreed the price.

DEPOSIT: Insist on cash; it concentrates the buyer's mind on your car. A cheque can be cancelled and a promise of payment isn't always fulfilled. A customer with second thoughts, or suddenly finding another car that takes their fancy, can soon fail to come up with the readies. If there is no deposit, there is no intention to purchase and no good faith. Specify that it is non-refundable. Don't believe someone who says that don't have any cash on them, as no one goes to buy a car without money. They have to give you cash or the car remains for sale. Simple. We have also mentioned digital, bank transfer and cryptocurrency options see onward.

RECEIPT: To acknowledge the money you should give the buyer a receipt and whilst you are at it, put the details of the deal on paper. Keep it simple, write down the facts about the car (registration, colour, make), the parties, (you and the buyer), the sale price and amount of deposit. The BanGlossary should help.

TERMS: The important part is for the car to be 'sold as seen', or 'sold as seen tried and approved' as you are not guaranteeing that it will be fault free for two years or anything stupid like that. Should the car be sold for spares, again make this clear as you could be prosecuted for selling an unroadworthy car. Any conditions that you have agreed should also be included. So if you've put the onus on the buyer to take the car the next day and pay in full, put this in the contract. If you've agreed to have the car tuned, again be specific otherwise the buyer could ask for a full service. Then both parties sign. Make sure you have a copy.

DON'T: This seems obvious, but don't hand over any documentation, or the car, until you have received full payment. Some unsavoury characters have asked to borrow registration documents when in fact they want to use it to sell a stolen vehicle of the same make. Don't keep all the documents in the glove compartment, or show the buyer where you keep them. Never let someone take the car away 'to show their mum', even though they don't have the registration documents it won't stop them selling it elsewhere, or nicking all the best bits.

PAYMENT - CASH: The best way to get paid, especially if it is a relatively small amount. Always count the amount, as buyers make mistakes, not always innocently, and get someone else to double-check it, especially if the denominations are small. If you are nervous about taking money arrange to take the money at a bank. The cashier will count it and be able to spot a forgery very quickly. Also some nasty buyers just want to show you the money and take the car with a bit of violence. So a bank is a safe place to do the exchange. **BANKER'S DRAFT:** Effectively this is the bank's own cheque and is as good as cash and is the only other fool/ full proof way of accepting payment on the day of delivery. There are though forgeries and when in doubt telephone the Bank. **CHEQUE:** Never take a personal cheque, unless you clear it first (usually three to five working days), or pay for special clearance. Actually who in their right minds would use a cheque? That's ancient technology, everyone can now do pretty instant **BANK TRANSFER,** a BACs payment is really easy and your bank will give you a quick notification on your mobile phone. Also free to do. ONLINE TRANSFERS There are also other transfers and Pay Pal is the best known but there are others. **DIGITAL/CRYPTO CURRENCY** This is a thing and an alternative to the usual Fiat currency (that's printed government money). I won't give financial advice, do your reaserch, BITCOIN is the best known digital, but there are tons of others. See the BANGLOSSARY chapter for a sample sale/deposit agreement. To sum up, get cash, but again money can be forged. Always find out from your government what the latest guidelines are to spot a forgery in your currency. The quickest and safest way is a BACs transfer.

Alternatively....

SELLING TO THE CAR TRADE There are dealers seemingly desperate to buy your old car, but the problem is they want to pay a 'trade' price. The advantage is that you get a quick decision and instant cash without the hassle of trying to sell privately. The normal rules apply so that you should clean and prepare the car and have all the documentation ready.

Part Exchange. Here is the most hassle free way of completing the transaction, which also gets you out of the bother of selling before you buy. The downside is that you will get the trade value. That price for your old Banger can be disguised by the dealer taking money from the profit they make on the new. If none of that matters to you this is always the simplest way to get rid of your old car. Check online classified advertisements finding out what Bangers similar to yours (age, mileage and model) are being sold for by other sellers. That gives you an idea of what to list your Banger for privately. A lower than average asking will get more enquiries, a higher price will make more money especially if the Banger is in good condition and has a decent history. Don't look in a car price book. These are usually called price guides and use the word 'guide' for a very good reason. It is guide price based on a vehicle that may be similar to yours, but can still be affected by a large number of factors.

AUCTIONS: Just as hassle free, but again cars are usually sold at 'trade prices' or less. You will also pay an entry fee and a percentage on the sale price. If you are lucky a bidder could possibly pay more than it is worth, but don't bank on it. There are of course online auctions, see back for how to buy which will also give you guidance on how to sell.

GIVE YOUR BANGER AWAY. Seriously, help someone out who desperately needs some wheels. If you have got full value out of your Banger, do some good and spread some Bangernomic Karma around.

SCRAP If your car has been pronounced unroadworthy or has died mechanically you may get the scrap value, or even a token amount depending on the value of metal at the time. Some specialist breakers yards may pay extra for an obscure or rare model. Also contact owner's clubs who may be able to make use of the parts and may take the car away for free. Ensure though they will dispose of the car responsibly, that may seem difficult, but maybe ask for proof (a receipt) that it has been done, otherwise the car could be traced back to you and you could be made liable for any laws that have been broken.

INCREDIBLE HULK Potential buyers have sniggered, a dealer offered a derisory sum and the neighbours have got up a petition for its removal. Your pride and joy isn't going anywhere, so what do you do? The simple answer may be to chop it into bits. But instead of giving those bits to the dustmen you can actually sell them off individually. The fact is that most bangers are worth more in bits than they are in one piece. OPTION 1: If you have the space, store the vehicle away and remove the pieces as you sell them. OPTION 2; Remove the most saleable items from the car and store those away for sale. One reason for keeping the old Banger is if you buy an almost identical replacement, because then you have your own parts store for free. Just make sure that the car isn't an unsightly hulk, rotting in your garden.

WHAT YOU NEED: Most basic tool kits have the necessary items to reduce a vehicle to its constituent parts: spanners; a socket set; pliers; wrenches; cutters; a hammer; chisels; hacksaw; releasing fluid/penetrating oil; and don't forget to don old clothes. Power tools can sometimes speed up the process, but they are not strictly necessary. A workshop manual, could also save you inflicting unnecessary damage on a part you want to salvage, but we told you to get one of those ages ago. Dispose of all fluids and other unsold parts responsibly, as a recycling yard will be pleased to take metal and old oil.

Bye Bye Banger – What can you do?

You know a lot about buying cars so put that to good use when it comes to selling. All those things you didn't enjoy about the buying experience you should try and put right. Write down how you could make the disposal/selling process easier for a potential buyer. Maybe you should not deal with the buyers because you haven't got the patience. Perhaps you will get your local garage to agree to give a report on the car's roadworthiness. Maybe you but together a highly detailed photo album or video film. I Dunno, anything unusual could help you sell, try and be different.

BANGERNOMICS
Breaking up is hard to do. How to dispose of your Banger, piece by piece...
Engine: Only worth heaving out if running well. Also most buyers prefer to hear it running before buying.
Banger: Whatever is left of it, there is always an intrinsic scrap value. No need to pay for it to be towed away.
Battery: If new with a receipt, worth reselling otherwise for use as a spare. Always has a scrap value.
Engine Parts: pumps, electrics radiator and so on.
Bodywork: Doors, bonnets, boot lids, and wings are saleable in good condition, or if rare.
Lights: These clusters if working and not cracked are always worth something.
Don't: dismantle in the street if you can avoid it and want good neighbour relations. Do: Ask buyer to remove parts.
Interior: Parts are hard to find, so there is a demand for seats and trim, can take a while to sell.
Wheels/Tyres: Keep them on so that the Banger remains moveable.

BANGERNOMICS
Part Exchange
Break for Parts
When selling keep control, have someone with you be safe and don't act desperate
Walking away
You could simply give your Banger away...
Buy Cheap
Buy Smart
Recycle

13. How to do Bangernomics

A. Basic

Any idiot can buy a Banger for a few quid, that's the easy bit. Much harder is finding a motor that works. The last thing you want is a monument outside your house as a constant reminder of just how rubbish older Bangers can be and of your stupidity.

There is no point buying a dirt-cheap Banger, then servicing it yourself when you may not have the time or the expertise to do this. All you need is a Banger that will fire up, first time, every time, not cost a fortune to run and that doesn't need to be even slightly interesting. What we have here then is the closest anyone will ever get to just buying a used car. Bangernomics is as simple as that, but the really important element is to make sure that you are covered. So you don't take chances. This then should be one of the shortest chapters in the whole Diet.

First of all, have a decent budget. Much less than a new car one so you are already up on the deal. It must be what you can afford, no borrowing or easy terms. Set a realistic, affordable amount and stick to it.

Only look at Bangers you know. Stick to the big brands like Ford, Vauxhall and Kia for example. Mainstream. Nothing premium (Mercedes, BMW, Audi), which can be priced accordingly, what you want is a value for money motor. Something others may have avoided because the colour is odd, or the specification is missing a few vital things, like air con, or sat nav.

Just in case you wondered, there is no reason to either be, or get a mechanic to check out your potential Banger. However, there is a soon to arrive chapter of the book that runs through the 'what to look fors?' The basics are, make sure there is a fresh 12 months of MoT certificate (or annual safety check, wherever you live). Also, don't imagine that you will be over paying, or getting ripped off when visiting a dealer. This is a point that has to be made quite strongly as it may come as a surprise to some.

Buy from a dealer because that means you will have the law and a warranty on your side. In short, there is some legal redress, so the dealer will have to sort out any problem. Contrast that with a private sale where it is very difficult if there is a dispute to get anything fixed unless you can prove fraud, or that you were misled. In most cases you will have

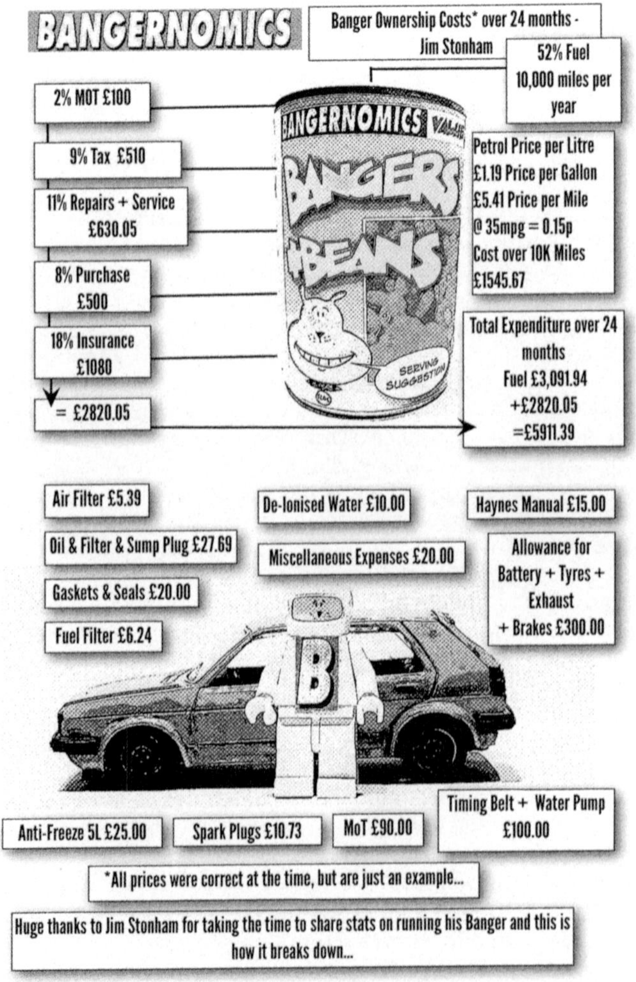

to deal with whatever issue your newly acquired Banger has. Then go after the seller for money and good luck with that. However, if you are offered a Banger privately and that has a continuous run of family owners with lots of history, plus it looks tidy, then maybe you should make an exception.

Dealers and car traders often get part exchanges,

which don't fit their stocking policy because it is the wrong sort of model. Perhaps the mileage, colour or specifications is wrong for them. Sometimes they advertise these, but it is probably just worth asking a local dealer if they have anything which is taking up precious space that they would like to move on. Don't worry if they won't give you a warranty that doesn't matter, you can go and buy one for some protection privately.

Warranties have a poor reputation and often that is correct, as some don't cover every eventuality. Actually they can be effective. Here is what you need to know.

A warranty is essentially an insurance policy, which covers your car against specific mechanical failures. Most new cars have a three-year manufacturer backed warranty. Buying a used car from a main dealer usually means there will be a 12-month manufacturer guarantee. Cars purchased from independent dealers and car supermarkets may be offered with a warranty package. It is also possible to buy your own warranty cover. There are plenty to choose from and some cover more than a simple mechanical failure. You can even get a policy that operates when a car fails its MoT and there are also packages, which includes maintenance and spreads any servicing and repair costs over the year. It will suit some drivers perfectly and remove any worries they have about unexpected expenditure.

To get the best warranty that suits you it is important to read the small print to see exactly what is covered. You may have a claims limit and may even have to contribute to any costs. Indeed high mileage and older Bangers may not be eligible for a warranty, usually if they are over a decade old or have covered over 100,000 miles. At the other end of the car market, some sports cars and complicated luxury vehicles may also be specifically excluded. Otherwise a good warranty will cover the engine, fuel and ignition systems, cooling system, electrical system, gearbox, transmission system, steering, suspension and selected clutch and brake system parts but not the actual wear items like the clutch itself or brake pads.

'Wear and Tear' is something that some warranty companies rely on this as their get out clause. When wear and tear is an issue then a good company will request a percentage contribution to the parts and labour costs. Most warranties exclude normal wear and tear to wheels, tyres, exhausts, catalytic converter, clutch and brakes, as well as damage to bodywork, paintwork, glass or headlights. There are also exclusions for certain electronic or electrical items such as CD, sound system and Bluetooth systems. Air conditioning and climate control can be excluded, but really those are effectively human rights these days and if a company does not cover those items, what else won't they cover? Best to avoid those policies.

A warranty is only likely to be valid if the vehicle covered is looked after properly. This means that in some cases an approved dealer must service the car in accordance with the manufacturer's schedule. That also means that the car must not miss its due servicing date by a certain number of days or miles. The obligation is on the owner to keep careful records and receipts in order to make a claim. With older Bangers though it is important to have some flexibility as you may not want to take a five year old car to a main agent, when an independent garage has significantly lower labour rates.

The total of all claims can be limited to the value of the car or a specific limit for each item. That limit may be too low especially if your car is usually expensive to repair. So it is important to check limits for individual claims and also the total claim limit in any one year. The warranty company may also have a stipulation as to the maximum labour rate they are prepared to pay for repairs. Plus if a new part is fitted which is better than the old one (upgraded engine part), this is referred to as 'Betterment' and will mean that you are expected to make a contribution to its cost. A good company will have a policy on consequential loss, where an excluded part damages a covered one so that you don't lose out.

There are companies that will for an annual fee look after your Banger when it comes to servicing, MoT and repairs and wrap all of that up with a decent vehicle recovery service so that you will never be stranded if it breaks down. This is a hands off approach to Bangernomics and there is nothing wrong with that. You just have to find the right one.

B.Hardcore Bangernomics

So far, we haven't been very hardline about what you should and shouldn't do to slash your motoring costs. For some, Bangernomics really is a brutal way of life. No Classics, or Luxo Bangers, just the cheapest Banger possible. Some of you could find all this a bit too intense, but the simple truth is there are a few tips and wrinkles that could help us all to own and drive more cheaply. Here we are just scraping the rusty surface of just how far the committed Bangernomicist will go to save money. How to get more useful miles out of an overpriced gallon by Hypermiling. A

BANGERNOMICS
Hardcore Version
Hardcore Bangernomicists will fix this Banger themselves, or chuck it away
Want most Banger for minimum price
Buy a Write-off
Shop at an Auction
Go for Max MPG

few simple changes in driving style and how a vehicle is used will boost mpg, better known as miles per gallon. Increasing the number of miles you cover to the amount of fuel put in the tank will bring huge savings over the course of a year. Drive smoothly to get more miles per gallon from a Banger. Sudden acceleration, rapid braking and not looking ahead to what the traffic is doing costs fuel. Lose weight to get more miles per gallon from a Banger. It is quite simple, the more weight a Banger has to move around the more fuel it uses. It is well worth emptying the boot of anything that isn't needed. Bags, clothes, maybe shopping you have forgotten about. It is daft to leave a roof rack, top box, or trailer attached if there is no need for their carrying capacity. It is a drag and slows you down which all increases mpg. The proper Banger Hardcorists might be tempted to take all the trim, carpets and unused seats out of a Banger. I mean that really is weight saving. You can disappear down a rabbit hole called Hypermiling. This is extreme fuel saving. We won't go there, but here's a tad more on making a gallon go further.

Despite the modern miracle of Sat Navs it is not always the answer. It may not take you the most direct route, so it is always worth checking on a good old-fashioned map and getting precise directions for the last few hundred yards of the journey. Driving around looking for the final destination wastes time and fuel.

Adjust tyre pressures to get more miles per gallon from a Banger. Tyres are crucial to maximising miles per gallon. Regularly check the manufacturer's recommended pressure levels and don't just inflate and forget. Low pressure costs fuel, high pressure makes the wheel roll better but wears down the tyre. That's why the recommended pressure is best.

Turn off the engine to get more miles per gallon from a Banger. Some Bangers may now have stop/start technology and there is a good reason for that, as it will save fuel when you are at a busy junction or traffic lights. Otherwise, when it is safe to do so, turning off the engine means that you will not be doing zero miles to the gallon. It never makes sense to be parked with the engine running. Using gadgets costs fuel and reduces miles per gallon from your Banger. The sad truth is that switching on the climate control costs fuel, as does listening to music or the radio. It might seem silly to swelter whilst listening to the sound of silence, but it might help squeeze a few more miles out of the fuel tank when the red light is on. Service your Banger to get more miles per gallon. It does not pay to avoid those manufacturer recommended visits to the garage. It will keep your Banger running at the optimum efficiency and that means getting the maximum miles per gallon.

A hardcore Bangernomicist may go to a car auction. This is not recommended for those who want an easy and stress free buying time. For those who want to buy Bangers the way some car dealers do, it has to be a car auction. There are lots more risks and no ways to reject a car once it has been bought. No test drives, or detailed inspection. Also many Bangers go to auction for a reason. Maybe it has an incurable misfire or a rumble at 50mph that the dealer can't sort out. Perhaps there is no V5 registration document, it can take weeks to sort out, sometimes months. Anyway, we will go there and do that, it is flipping scary and dangerous. Best left to the hardcorists...

Hardcore Bangernomicists may buy an insurance write off. According to the vehicle data provider HPI, one in twenty five vehicles they check are recorded as insurance write-offs and insurers are writing off over 500,000 cars a year. Category A is Scrap, B is for Break, neither of those categories can be returned to the road. Category S has structural damage (where the car's frame or underpinnings are damaged). N is Repairable Non-Structural. Vehicles in Category N is where the cost of repair exceeds the car's value but there is no structural damage, like a dented wing. What you have to bear in mind is that a vehicle which is on the damage register may be cheaper, but it will always be worth less and more difficult to resell.

The Hardcore Bangernomicist won't be reading this, because they would never buy or feel the need to have a Bangernomics Book in their lives. Mind you, if someone showed them these paltry few pages, they'd be hugely disappointed. So don't show it to them, let 'em bore you rigid in the corner of a car park about just how committed they are to saving a small fortune on their running costs. Listen, because you might learn something. It may also be difficult to walk away.

14. THE TEN BANGERNOMICS COMMANDMENTS

Just as Bangernomics is not actually a proper Diet, there are no proper rules. What we have here are a set of guidelines, some recaps of what it is all about and a few motivational words. If you would like to interpret these as rules then feel free to break, bend and ignore them. Simply go with your gut feeling. If the Banger is within your budget and you like it, buy it. Having said that, it is much nicer to have a neat finish, which ties everything together. So are there really any hard and fast rules when it comes to buying and running a used car? Sort of. Even if you ignore one or two then the others should help you to wriggle through the dangers and get a great Banger.

1. A year's Ticket. Always make it a condition of your purchase that your Banger passes the MoT. In the old days it was referred to as a 'ticket' in old car dealer slang and a full one is a very good thing. A roadworthy Banger means that you will have a good start to Bangernomic life. It is not a 100% guarantee that nothing will go wrong as it only applies at the time of the test, but at least you will know what is right and what may go wrong in the future, because of any advisories

2. Tyres maketh the Motor. Yes we are obsessed by tyres, but quite right too. They are what keep you on the road at all times of the year. To be safe and not sorry it is essential that you avoid anything, which is either a Remould (basically a remade old tyre), Retread, pretty much the same, or a mix of makes and treads which all spell serial neglect. Avoid if you cannot establish where the tyre comes from. Worth doing a bit of research on those precious donuts.

3. Broaden your Buying Mind. Everything is up for grabs. If there is an opportunity to buy a Banger Tractor, then do it. Essentially there is no point in being precious or deciding on one make or model, be open minded and get something that you can afford and fits your lifestyle. That's Bangernomics.

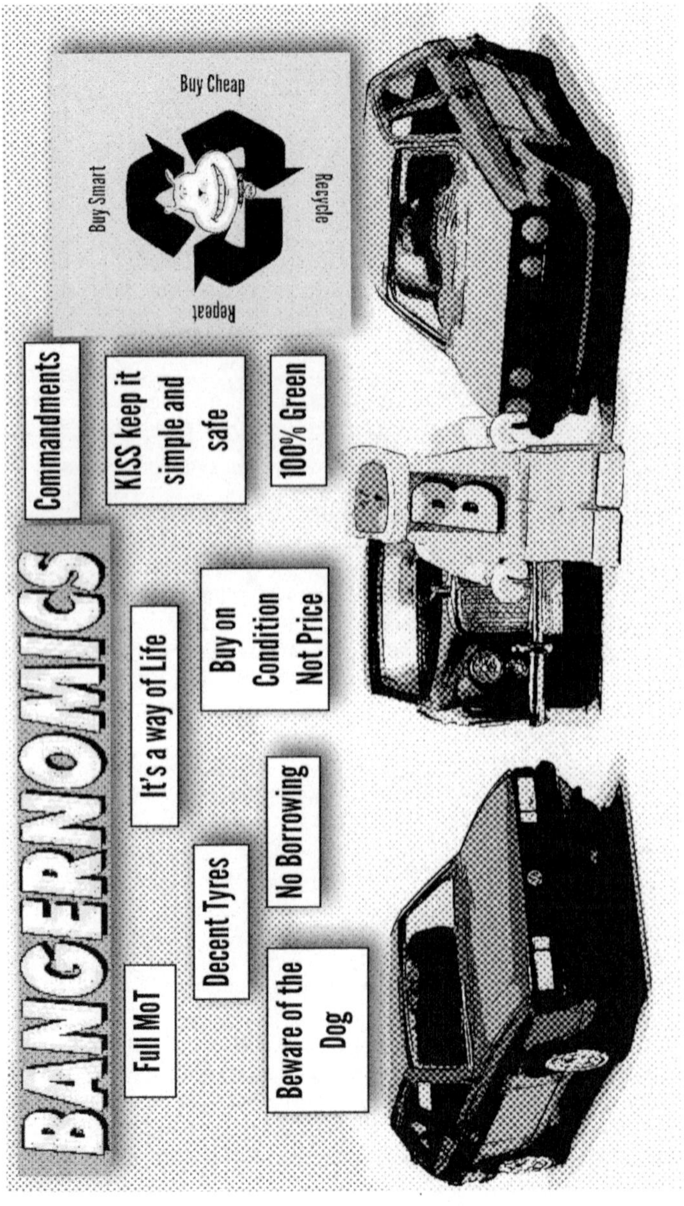
BANGERNOMICS
Commandments
Full MoT
It's a way of Life
KISS keep it simple and safe
Decent Tyres
Buy on Condition Not Price
Beware of the Dog
No Borrowing
100% Green
Buy Smart
Buy Cheap
Recycle
Repeat

4. Don't Borrow. Always pay cash, that's real money and never borrow to finance the Banger, that way you will be spending what you can actually afford.

5. Bangernomics is all about commonsense If it goes crash, bang wallop, looks wonky and the person selling it you wouldn't want to touch with two barge poles lashed together, that usually means: best not buy it. The Bangernomicist looks and listens carefully and is never rushed into making a hasty decision.

6. Keeping it Simple and Safe. That is what Bangernomics actually does. You can complicate it, but best to buy the simplest Banger you can that needs the least amount of attention. That is the dream anyway, but a reliable, practical Banger will transform your life. Oh yes, don't worry that a Banger does not have Driver Assist somesuch, a Lane Change Beeper, or Bouncy Castle sized airbags. Owning a more mature car will make you a better road user. Drive defensively, be aware of your surroundings and maintain your Banger properly and you will be doing so much for overall road safety as a result. With the money you save, take advanced driving lessons.

7. Buy on Condition, never price. Even if you are offered a free Banger, it is all about condition. The trouble is that sorting out a Banger with problems, can be very problematical. More than that, it will cost you real money.

8. Bangernomics is Green. (Yes I know we have covered this but it bears hammering home) According to some Eco fascists car ownership is irresponsible because of the pollution they cause. However, a Bangernomic approach amounts to recycling because the natural resources and energy used in building a new car is phenomenal. So prolonging the life of and disposing responsibly of a used car is very green indeed. Why throw away a car that still works? You wouldn't demolish a house just because you moved to another one. If there is life left in it, sell or give it to someone else. Bangernomics is recycling an entire car – brilliant.

9. Bangernomics is a way of life. Not only is Bangernomics easy, it is mostly fun. Changing oil, plugs and filters is within all our capabilities given a secondhand owner's manual.

However, finding a good local mechanic rather than paying through the nose for a franchised dealer charging loads an hour is the key if you don't fancy getting your hands dirty. Meanwhile the MoT helps to ensure that the car is safe. Yes you learn some skills, waste less money and have a new topic of dinner party conversation, because Bangernomics is a way of life.

10. Always beware of the Dog There is no excuse for buying an unroadworthy heap and Bangernomics helps reduce the risk of that. However, Banger buying is always going to have a large percentage of luck. At best there could be a 50/50 chance that you could end up with a pup. However, follow the guidelines in this Diet and you should find that the odds are stacked in your favour.

To sum up then, the Bangernomics Diet is one way to buy a Banger and save lots of Beans. The Bangernomics Bible was an update of Bangernomics and Bangernomics Rebooted is just an out of date reprint.

Bangernomics is an alternative motoring manifesto for the masses.

15. A to B to Z of Bangers:
Which really means contact Bangernomics.com for Free Lifetime Support, are we mad?

Just about here in the original Bangernomics book I used to list loads of cars from the '60s onwards and although I tried really hard it was never going to be a comprehensive listing of all past, present and potential Bangers. I may not have succeeded and possibly started a few arguments. There is now of course the Interweb, which has a huge network of owner's clubs and forums that can provide inside information on every make and model of car ever. Then again I have driven myself mad by writing Ruppert's Bangerpedia which I can highly recommend. Contains over 700 models made from 2000 to 2010.

My own prejudiced, often juvenile opinions are based on a lifetime of mixed motoring experiences, driving disappointments and numerous ill-advised purchases and sales. All I can do as an act of Bangernomic good faith is to provide you with lifetime support. Initially it will be my lifetime, however long that turns out to be, although I will try and sort something out in the far flung future. So while emails and the Internet exists you can contact me (for as long I exist) for biased advice through the Bangernomics.com website, or whatever clever interactive technology replaces it.

16.Banglossary: *Checklists etc*

Here they are, all the crib sheets, checklists and standard forms you will need to help you become fully qualified in the science of Bangernomics. There may be more information on it than you need, but it is a fairly simple checklist which will help you calculate quickly whether the Banger you want to buy is a good one. So if you are applying more crosses than ticks, it might be worth looking at another Banger.

The point is use them as reminders of what to look for, you can write in the book, but after a couple of Bangers it could become rather confusing.

Don't get too hung up on testing everything, especially if you are not sure what an exhaust is and exactly where the petrol tank may be. If something looks wrong then make a note for yourself and ask an expert later.

Because we are so generous the author and publishers (which is basically just me) grant permission to the purchaser of this book/course to make copies of these lists for your own private Bangernomic buying purposes. Please don't copy them to resell or redistribute and ideally if you have a friend who is interested in Bangernomics, ask them to buy the book. I'd really appreciate that.

I am obliged to issue a disclaimer that these lists and forms are not comprehensive and the royal we take no responsibility for how you use these materials and we can't guarantee that if you use these lists you will buy a brilliant Banger. However, we do think that these will probably help, a bit.

GENERAL GUIDELINES

ACCIDENT? If body panels have been repaired, the paintwork looks patchy and has different textures, you can see repairs or welds in the engine bay, boot area or underneath.

FAKED IDENTITY? VIN plate fitted in engine bay and on dashboard tampered with, windows and lights and all glass with scratched off areas, stickers over etching, documentation does not match car.

CLOCKED? Mileometer figures out of line, condition inconsistent with indicated mileage, mileage on service documents very different from indicated mileage,

documentation missing. High mileage not necessarily a problem, but if the seller has lied about this, or altered the mileage in any way, what else have they changed on the car without telling you?
STOLEN? Documents inconsistent with owner or car details, registration plate changed, VIN plates tampered with, documents do not look genuine.

VEHICLE DETAILS
Get these from the car itself so that you can compare them with the documentation supplied with the car and what is online.

Make/Model______________________
Year______________________
Colour ______________________
Registration______________________
VIN______________________
Engine No ______________________
Mileage______________________

DOCUMENT CHECK Take your time and carefully look at all the paperwork, if any is missing ask where it is.

MOT/Annual...Check/Roadworthiness
Certificate__________________
Does it look genuine, does it match other details on the car?

Registration Document______________
Does it look genuine, does it match other details on the car?

Warranties____________________
Are there any relating to tyres, the car etc? Ask.
Tax_________________________
check online

Insurance____________________
Are there any documents, do they match the car/owner?

Last Service____________________
Is there is any service history?

If you use chalk

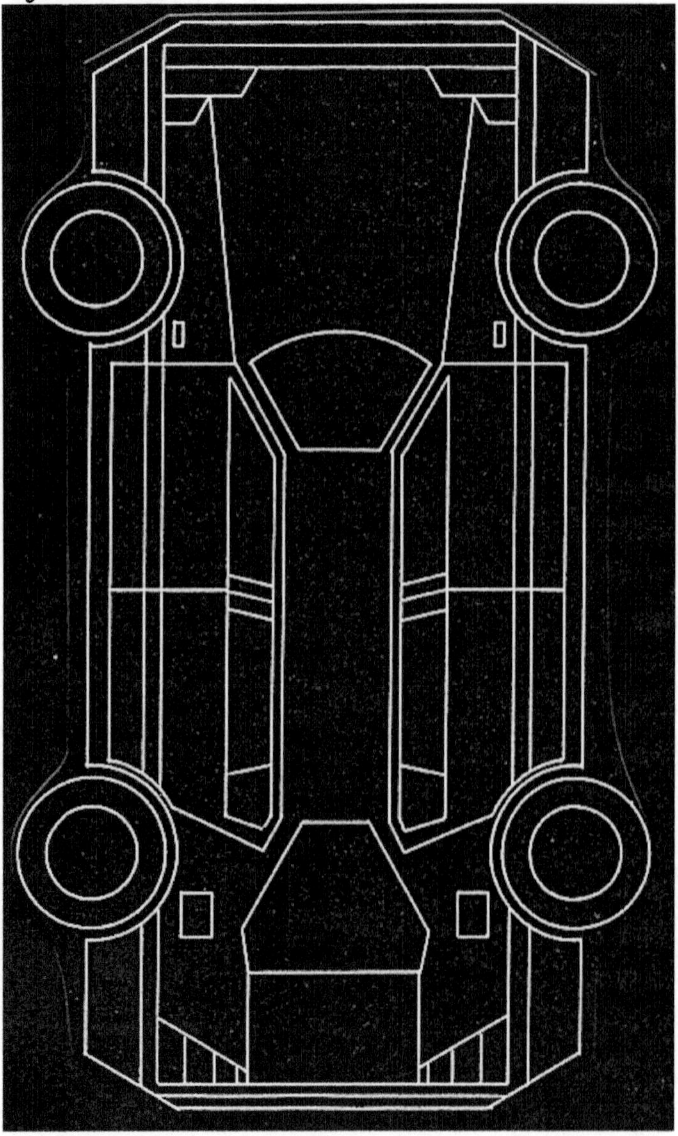

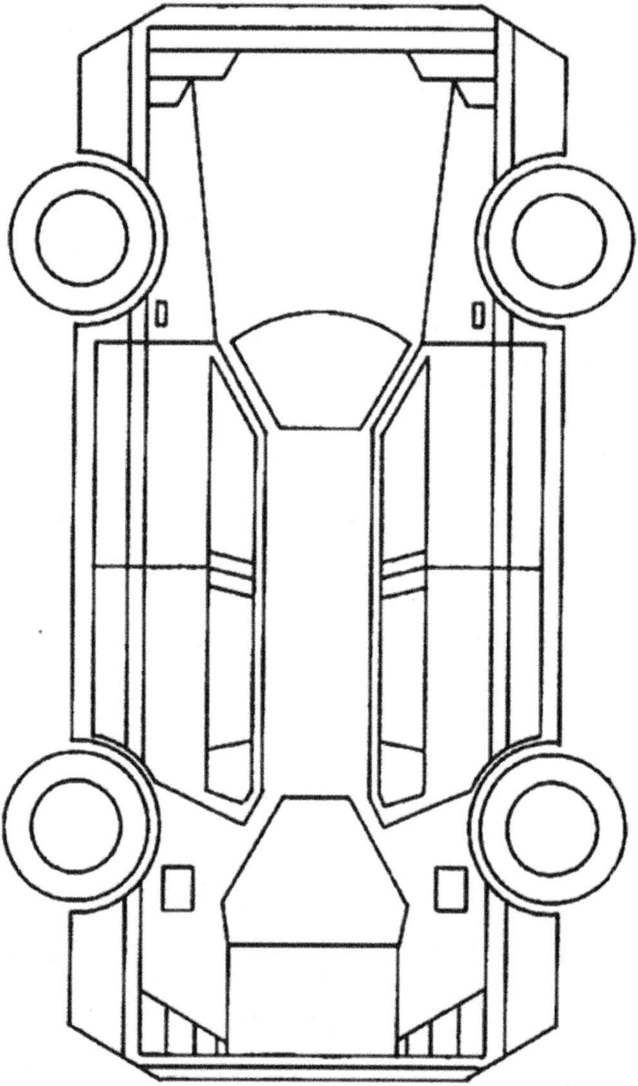

BODY CHECK Use the flat drawing of the car to identify any damage and also this key will help. Tick or cross as appropriate.and then make any comments
key F = front **R** = rear **D** = damage **C** = corrosion **O** or circle = repairs Use a tick to indicate that the item is satisfactory an **X** to suggest there is a problem.

door locks
door mirrors
exterior trim
fuel filler cover/ cap
glass
number plates
paintwork
soft top operation
WHEEL/TYRE CHECK
tyre front left
tyre front right
tyre rear left
tyre rear right
wheel rims
wheel trims
LIGHTS/WIPERS CHECK
(You will need a friend to
help you with the lights)
auxiliary lights
headlamp washer
headlamp wiper
headlights
indicators/hazards
number plate light
rear lights
reverse/fog lights
side lights
stop lights
wipers/washers
ENGINE CHECK When cold.
bonnet operation
coolant level
drive belts
engine oil level
brake fluid level
clutch fluid
hoses/pipes

cold starting
fumes/smoke
noise level (cold)
oil leaks
steering fluid

INTERIOR CHECKS Use the interior plan.
adjustable mirrors
air conditioning
boot lock
carpets
central locking
cigarette lighter /socket
door locking
door seals/hinges
door trim
headlining
heater fan/controls
ignition lock
illumination lights
interior sills
luggage area
mirrors
panel lights
rear parcel shelf
seat belts
seat mechanism
seat trim F
seat trim R
sound system
spare wheel
steering adjustment
sunroof
switches/controls
tool kit
windows

UNDERSIDE CHECK Don't get underneath, don't use a jack, just see what you can from just getting on the ground. Use the underside picture. Look for any damage, obvious leaks and corrosion that you should mark with a (C). This is a visual check of course and not a mechanical assessment.
oil leaks
pipes/hoses
suspension

EXHAUST CHECK
joints/couplings
mountings
pipes
silencer(s)

TRANSMISSION CHECK
fluid/oil leaks
mountings

FUEL CHECK
fuel lines
leaks
tank

BRAKE CHECK
discs/pads
flexible hoses
fluid leaks
pedal/linkage
pipes/connections

TEST DRIVE CHECK Make sure you are insured and bear the following points in mind. The car should start easily. The oil and ignition light should go out quickly. It should pull away smoothly without any strange noises, it should not jump out of gear or perform sluggishly. On a hill the car should not jump out of gear. Pull away on a hill, the handbrake should hold the car and the clutch if the car is a manual it should hold the car on the hill. On the descent there should be no blue smoke. If you can, try an emergency stop on a clear quite road. The car should not pull to one side and the brakes should not squeal. The steering should not wander if you release your grip on the steering wheel. Over rough ground the car should not pitch too much and listen out for banging noises. When cornering there should be no knocks, clonks or clicks, or general rumbling from the wheels. On a dual carriageway listen out for and feel for vibration. An automatic gearbox should change smoothly.

4wd operation
clutch
engine noise

engine
excess smoke
footbrake
gearbox
hand/park brake
hot restarting
controls
overheating
road holding
steering
suspension noise
warning lights

AFTER THE DRIVE: With the engine running look for oil leaks in the engine bay any leaks from the hoses and smoke coming from the oil filter if you can see it.

If you get as far as actually buying then here is a receipt and holding deposit templates you can use and adapt. As ever this is not a legally binding document, but it may help you collate evidence if there is a dispute.

BANGERNOMICS RECEIPT
I...
(SELLERS NAME) of...
....
(ADDRESS) confirm that l am the owner of the vehicle
MAKE.,........
....
MODEL........
....
......
REGISTRATION.......................(or that I am authorised to sell it on behalf of the owner) NAME........................
....
ADDRESS:.......................
....

I confirm that the vehicle does not have any faults which I have not disclosed/and that the recorded mileage of.............. ......is correct. I agree to:
....
....

.... before sale.
(DETAIL REPAIRS)

HOLDING DEPOSIT I ____________ (SELLERS NAME) accept the sum of £........
as a holding deposit on the above vehicle, which is fully returnable provided a decision is made within ___days/ hours. The purchase price for the above vehicle will be £................
SIGNED...
DATE........................

CHARLES WARE'S
MORRIS MINOR CENTRE
0117 3003754
XWV 452A

17. CHARLES WARE INTERVIEW *The Man who saved the Morris Minor*

The late Charles Ware, championed the Morris Minor as a durable, economic and loveable alternative to a new car every three years. Not only that, in 1991 he established a plant in Sri Lanka to make the complex rounded Minor body panels by hand. This is a proper partnership between Charles Ware's Morris Minor Centre and local interests in that country. So before it became fashionable to be Fair Trade, Mr. Ware was doing his bit. Not only that, owning a Minor or a simple classic Banger would also help local businesses to you rather than benefiting the global conglomerates that run car franchises and supply parts. Your friendly local garage can look after a Minor and maybe you could too in the spirit of self-sufficiency. Here is an interview I did with many years ago. He was charming, polite and incredibly knew who I was.

You might not have heard of Charles Ware, but his Morris Minor Centre (established 1976) is proof that sustainable car remanufacture and ownership has never gone out of fashion and could even be a way forward. Ware would prefer that we all think of car ownership in decades, rather than the short term approach that the car industry has lived and could soon die by.

As Ware shows me around the premises and the possibly the largest number of Minors in captivity, his affection for the cars is obvious as he touches each one in a reassuring manner, so why the Minor?

"It is fundamentally an over-engineered car." Ware grasps the metal and points out the heavy gauge material, taps at the substantial bulkheads and invites me to experience the solid thunk of the doors. There is no arguing that these are brick outhouse motors, especially as Ware's engineers can cut great chunks out of the structure without worrying about it bending out of shape. As for the old A series engine, "that will go on forever," he says. "One of my customers you could hear approaching from miles away, but she always resisted any engine work despite the obvious evidence that it needed an overhaul."

Another key ingredient to the Minor's longevity is the electrics, or rather lack of. Ware holds up four fingers, "It really is as simple as that, you can see what goes where, the fundamental flaw with modern cars is that the electronics now dictate the life of the car. It is planned obsolescence and we are now caught in a frightening cycle of waste." Ware admits that build quality is infinitely higher now and the virtual eradication of body rot is to be welcomed, but the wholesale reliance on electronics and culture of replacement rather than repair is killing modern cars."

Ware has always used the house analogy, which sums up his position perfectly. Essentially you would never demolish a perfectly decent solidly built Edwardian or Victorian residence just because a few tiles were lose or the central heating needed a new boiler. "New car buyers have been brainwashed into wanting the latest model and gadgets when they really don't need them."

Morris Minors are great cars. The light, direct steering and sporty handling are a revelation plus there is so much you can do make them more 21st Century friendly, from new seats, to electronic ignition, which Ware heartily approves of ("simple and reliable"). Servo brakes are a popular upgrade with Morris 1100 drums, Ford Sierra five speed gearbox makes longer journeys more comfortable. Nevertheless the hardcore of Ware's customers simply want, "a car that will get them from A to B reliably and cheaply, but they also want identity, something nostalgic they can treasure."

Surely that means that Ware's customers are a bunch of mean old duffers?

"Oh no, we have lots of people in their twenties and others who want a second car. Plenty of them are women who are not obsessed by speed and just love the shape. They also love the fact that the Minor is so easy to drive, the rack and pinion steering is so light and the whole chassis is set up for precise handling, it really is a joy.

"Over forty years new car buyers are going to lose a six-figure sum whereas rebuilding a Minor every few decades is

only gong to be a fraction of that, plus Ware has a common sense approach to restoring customers cars.

“The patchwork quilt attitude just won’t work and we will not do it. We always replace a series of panels that are crucial to the structure. A £50 weld at a local garage is pointless, far better to spend £200 to replace a whole panel that will still be there in ten years time.”

The problem is that a Minor which looks serviceable enough on the surface is actually rotten underneath and Ware says that no one takes the time to look at the chassis. Indeed, a customer that bought a Minor for £5000 online still needed £2000 spent in the Minor Centre workshop to make it completely sound.

Prices? (Rather out of date now) In theory you could turn up with a £300 Minor that The Charles Ware Morris Minor Centre will bring back to life for something like £10,000. That’s the cost of a fairly ordinary supermini at the moment. £15,000 gets an as new Minor which really will last you a lifetime. There are cheap Minors around, but they may not be the sort of Moggie you can rely on. Charles Ware knows his Minors and that means £3500 for a good saloon whereas Travellers start at £6500.

Let’s leave the final words to our hero Charles Ware who is keen to point out that his business is not some quirky automotive anomaly.

“People buy Minors not purely because it is retro, or because they want a slice of olde England, it is because the customers are normal, are from every strata of society and like the idea of a well-designed car that can be modified slightly to suit their needs. Fundamentally the Morris Minor is a brilliant design. All credit to Alec Issigonis. I like to think of the Minor as a room on wheels. Somewhere we would all like to be.”

James Ruppert Dreamt up the word Bangernomics in 1990 and wrote a story about buying and running an £80 Eastern Block Banger. Book of the word came out in 1993 and has found that banging on about marginal motoring only makes a marginal living. Please consider the following titles for inclusion on your bookshelves to help pay for food, fuel and automotive tat.

Order Books from www.bangernomics.com

There might also be T-Shirts, world shaped beachballs and old Daihatsu Applauses...

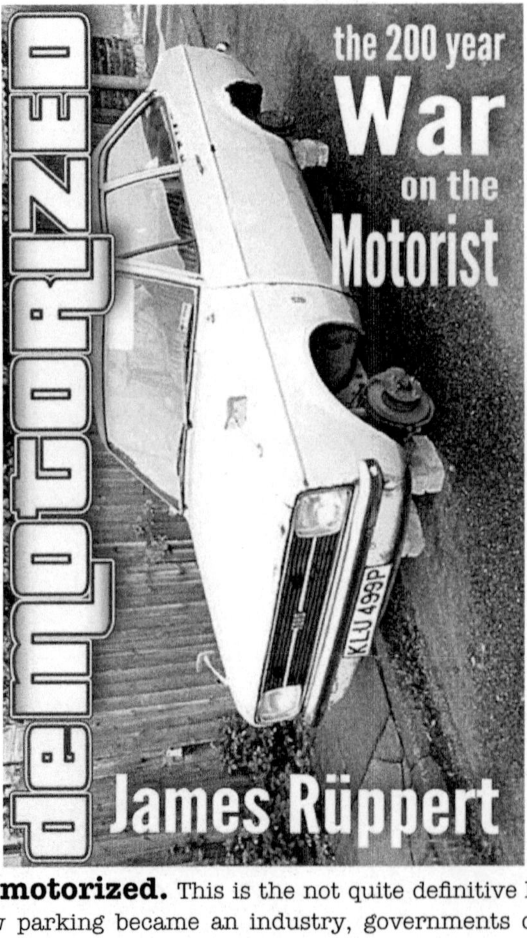

Demotorized. This is the not quite definitive history of how parking became an industry, governments overtaxed motorists, carmakers lied about how safe and environmentally friendly they were and tested their emissions on monkeys and humans. Also why robot cars are so dangerous and electric cars are not the answer. Also, who is behind the ongoing plan to Demotorize the world. Printed in 2020 it predicted the future with terrifying accuracy.
Buy it before it gets banned.

The Award Winning and so good the BBC based a documentary (Das Auto) on it **The German Car Industry, My Part in its Victory**. Winner of the Mercedes-Benz Montagu of Beaulieu Trophy for the best motoring book of year. Contains Yuppies, BMW 3, 5, 6 & 7-Series (yes that's all there was back then) and references to contemporary 1980s music.

The Critically acclaimed and best selling **The British Car Industry Our Part in Its Downfall**, which has been described as "Beautifully written with considerable wit" by the Daily Telegraph and The Daily Express said, "It's informative, well-written and extremely funny" which was really nice of them. The story of the motor industry from 1945 to 2005, explaining just why you can't buy a British built family hatchback from a British owned company. Plus all the cars, James's Dad bought.

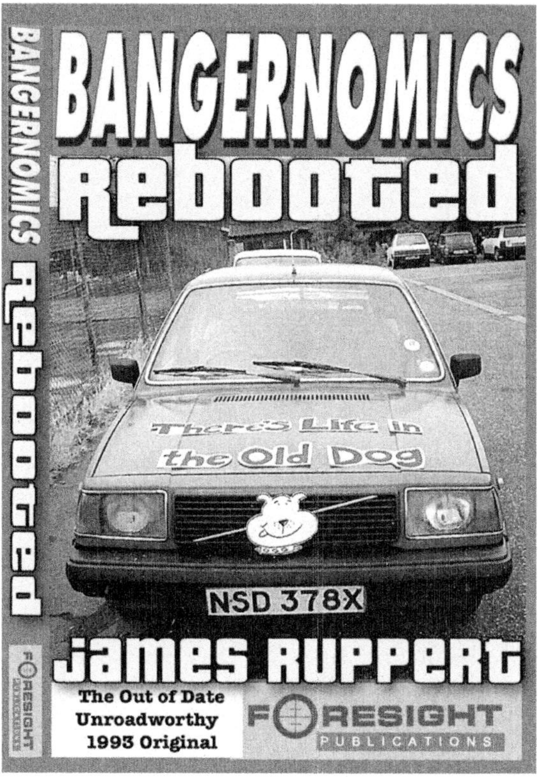

In the beginning, 1993, there was Bangernomics and now there is **Bangernomics Rebooted** which contrasts the absurd expense of buying a new car with the supreme good sense of buying well used. At a stroke depreciation no longer became an issue, running costs were slashed and there were no finance charges to be endured. Here is the original book, republished and effectively rebooted. So it is absolutely useless to any car buyer unless they have access to a time machine.

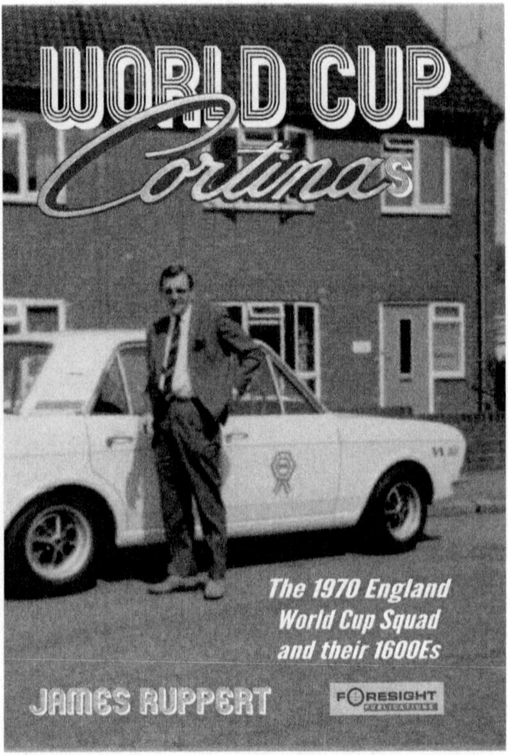

World Cup Cortinas In 1970 something very remarkable happened. Not only had England qualified for the World Cup in Mexico because they were defending champions, but most important of all Ford had loaned every squad member the greatest saloon car in the world. A book about what happened to those Ford Cortinas and how footballers bought better cars when they earned more money and no longer owned the same motors that your mum and dad drove.

“Channeling the absurdist spirit of Spike Milligan” said Classic Car, others just called it “Witty”, that’s **My Mini Cooper Its Part in My Breakdown** which has the full history of the world’s favourite small car, plus the history of what came before but not after. Also details the on-off restoration of a Mini Cooper that cost £200 in 1979. The expense of restoring the bodywork and mechanicals ought to put off most sane people. Except that original minis are always worth far more than you could imagine.

Buy real books, digital ones and T-shirts at
www.bangernomics.com

Ruppert's Bangerpedia deals with the confusing line up of noughties models. The Bangerpedia does this with a fairly ridiculous rating system involving Slog the Bangernomic Dog. The briefest of guides to each of the 700+ models includes some fairly marginal buying advice, a sarcastic appraisal and an almost definitive indication as to whether it will fit into a garage, or not.

Printed in Great Britain
by Amazon

82109550R00119